ORIGINAL PAPERS

BY

JOHN HOPKINSON

ORIGINAL PAPERS

BY THE LATE

JOHN HOPKINSON, D.Sc., F.R.S.

VOL. I.

TECHNICAL PAPERS

EDITED

WITH A MEMOIR

BY

B. HOPKINSON, B.Sc.

CAMBRIDGE:
AT THE UNIVERSITY PRESS.
1901

CAMBRIDGE
UNIVERSITY PRESS

University Printing House, Cambridge CB2 8BS, United Kingdom

Cambridge University Press is part of the University of Cambridge.

It furthers the University's mission by disseminating knowledge in the pursuit of education, learning and research at the highest international levels of excellence.

www.cambridge.org
Information on this title: www.cambridge.org/9781107455986

© Cambridge University Press 1901

This publication is in copyright. Subject to statutory exception and to the provisions of relevant collective licensing agreements, no reproduction of any part may take place without the written permission of Cambridge University Press.

First published 1901
First paperback edition 2014

A catalogue record for this publication is available from the British Library

ISBN 978-1-107-45598-6 Paperback

Cambridge University Press has no responsibility for the persistence or accuracy of URLs for external or third-party internet websites referred to in this publication, and does not guarantee that any content on such websites is, or will remain, accurate or appropriate.

CONTENTS OF VOL. I.

	PAGE
Memoir	i
Memorandum on Engineering Education	lxiii
(Written for a Cambridge University Syndicate in 1890.)	

1. Group-Flashing Lights 1
   (Pamphlet first published in October 1874.)

2. The Electric Lighthouses of Macquarie and of Tino . . 7
   (From the *Proceedings of the Institution of Civil Engineers*, Vol. LXXXVII. Session 1886—87, Part I.)

3. On Electric Lighting 32
   (From the *Proceedings of the Institution of Mechanical Engineers*, April 25th, 1879.) (First Paper.)

4. On Electric Lighting 47
   (From the *Proceedings of the Institution of Mechanical Engineers*, pp. 266—274. April 23rd, 1880.) (Second Paper.)

5. Some points in Electric Lighting 57
   (A Lecture delivered at the Institution of Civil Engineers, April 5th, 1883.)

6. Dynamo-Electric Machinery 84
   (From the *Philosophical Transactions of the Royal Society*, Part I., 1886.)

7. Dynamo-Electric Machinery 122
   (From the *Proceedings of the Royal Society*, Vol. LI., 1892.)

8. On the Theory of Alternating Currents, particularly in reference to Two Alternate-Current Machines connected to the same Circuit 133
   (From the *Proceedings of the Institution of Electrical Engineers*, pp. 3—21, November 13th, 1884.)

CONTENTS.

		PAGE
9.	Note on the Theory of the Alternate Current Dynamo (From the *Proceedings of the Royal Society*, Vol. XLII., pp. 167—170.)	152
10.	Alternate Current Dynamo-electric Machines (From the *Philosophical Transactions of the Royal Society*, 1896, pp. 229—252.)	156
11.	An Unnoticed Danger in certain Apparatus for Distribution of Electricity (From the *Philosophical Magazine*, September, 1885.)	184
12.	Induction Coils or Transformers (From the *Proceedings of the Royal Society*, February, 1887.)	188
13.	Report to the Westinghouse Company of the test of two 6,500-watt Westinghouse Transformers. May 31st, 1892	192
14.	Presidential Address to the Institution of Electrical Engineers, January 9th, 1890 (Magnetism.)	212
15.	Inaugural Address, Institution of Electrical Engineers, January 16th, 1896	236
16.	Presidential Address to the Junior Engineering Society*, November 4th, 1892, On the Cost of Electric Supply (From the *Transactions of the Junior Engineering Society*, Vol. III., Part I., pp. 1—14.)	254
17.	Relation of Mathematics to Engineering (From the *Proceedings of the Institution of Civil Engineers*, Vol. CXVIII., Session 1893—1894, Part IV. "James Forrest" Lecture.)	269

* Now the Institution of Junior Engineers.

PORTRAITS.

Dr Hopkinson at the age of forty-eight . . . . *Frontispiece*

Dr Hopkinson at the age of about twenty-three . . *To face page* xxx

EDITOR'S NOTE.

WITH the exception of the Memorandum on Engineering Education, the whole of the papers here reprinted have been published in various periodicals and books, and I have to thank the owners of the copyrights for permission to republish them. Dr Hopkinson issued a reprint of certain of his papers in 1893*, and intended to publish a complete collection. I have endeavoured in this reprint to carry out his intentions as to editing so far as they could be ascertained. In a few cases I have copies on which he noted certain omissions as desirable; in others I have had to use my own judgement as to what he would have done. The omissions on the whole, however, are few; and the papers for the most part are reprinted just as they appeared with obvious corrections. Vol. I., roughly speaking, contains the papers on technical subjects, and Vol. II. those of a more scientific character. The papers fall naturally into groups based on their subjects, and have been so arranged.

I have to thank Mr G. T. Walker, Fellow of Trinity College, Cambridge, for kindly reading the Mathematical papers at the end of Vol. II. for me.

* " Original Papers on Dynamo-Machinery and allied Subjects." Whittaker & Co.

MEMOIR.

THESE volumes contain all that John Hopkinson wrote of an original character on Engineering or Scientific subjects. They form a record of his career to which little can be added. His practical inventions, and certain parts of his constructive engineering, however, are not directly illustrated in them, and a short account of these may perhaps enhance the value of the book as a memorial. It is also desirable from this point of view, as well as interesting as a matter of history, to trace the influence of his work on the development of the Arts and Sciences with which it dealt. There have been few men in whose minds the spirit of scientific research has been so closely associated with the power of practical application as in Hopkinson's. His scientific work was always informed with a desire to satisfy practical needs, without being unduly cramped thereby. Thus, while in many cases it resulted in rapid Engineering advance, it also exercised a great influence on methods of research. It was conceived in sympathy with practical aims and practical men's ideas and yet it did not cease to be pure and disinterested seeking after knowledge. To those who know his work in only one of these aspects, a connected account of it showing something of the other may be acceptable; and finally, a little should be said of his education, which was in many ways remarkable.

To give much more than this in the way of biography is hardly possible. John Hopkinson's life had few features of interest for the general public: this is, perhaps, inevitable of one whose work dealt with natural and not with human forces.

In the relations with the world of a statesman or a soldier personality is all-important and feeling may be a very powerful factor. It is therefore of interest, after his death, to know the emotional as well as the rational side of his nature, and there is justification for revealing it by a detailed account of his private life. With an engineer it is different. If a man would command Nature, emotion is out of place; and the more he can efface himself and become a thinking machine, the more likely he is to succeed. Hopkinson had this power in an uncommon degree; once in his office, his reasoning faculty asserted itself to the exclusion of all others. It is true that he could command not only the respect, but the love, of those who served him, so that they worked for him as they would for few. But this power was accompanied by no feeling on his part, and, though it was doubtless due to a subtle influence which could not have been exercised by one of less perfect character, it may be said broadly that a very great part of his personality found no expression in his work and therefore does not concern the public. There seems, in his case, to be no excuse for revealing in its entirety after his death, what in life he did not reveal. He was the most reserved of men, and only his family knew him intimately.

Another cause would make it difficult—if it were desirable—to write an interesting life of John Hopkinson. It sometimes happens in a great man's career that the whole force of his nature is pitted in a struggle with adverse circumstances. Such occasions form the most valuable items in biography. But they depend for their interest on the sternness of the struggle no less than on the might of the combatants. It is something for the biographer to be able to say that his subject overcame such and such things: it is far more if it can be said that he only just overcame them. Of strife of this evenly-balanced kind there was none in John Hopkinson's life. He had his battles to fight like other men, and against odds that lesser minds could not have faced. But the issue never seemed in doubt; he always won with an unknown reserve in hand. From contemplation of what he did, one can infer no limitation to his powers. And in such limitation centres no little of the interest of biography, for upon it depends in large measure the sympathy which the reader may feel for the subject. In consequence, too, of his freedom from disappointment, Hopkinson changed remarkably little after early manhood. In most lives adversity comes in plenty and leaves its mark—good or bad— upon character: none came to John Hopkinson. For twenty-five years the only change in him was a gradual and unresisted expansion. As he came to feel his strength his power of expression grew, and with happy family life his sympathies widened: in other ways he altered little. Reading the letters which he wrote long ago from Cambridge and Birmingham, on the threshold of his career, one seems to see the same man in all essentials whom one knew quite recently. I feel, therefore, that it will suffice for the non-technical part of this introduction if I can give a short account of him as he was then, and of how he came to be so. This course is possible because biographical material is fairly plentiful during the first thirty years of his life. His letters reveal far more of himself then than later, when they came to be mere records of passing events, given without comment and affording but little clue to the state of mind of the writer.

John Hopkinson was by birth and breeding a North-countryman. His mother—born Alice Dewhurst of Skipton—was of a family which has occupied a leading position in that town for several generations. Her father was the founder, and her brothers for long the heads, of the successful cotton-spinning factory which bears their name. Mr Dewhurst, John's grandfather, was in many ways an ancestor to be proud of. In early life he suffered much trouble in the death of his wife and all his family; but he married again, and had many children. One can infer that he possessed much of the immense vitality and energy which distinguished his grandson. As John knew him, he was a fine old Yorkshireman, impetuous, even fiery, but possessed of the kindest of hearts. He was much troubled with rheumatism, but he fought his malady with the most indomitable courage and tenacity. His grandson had a vivid recollection of his hobbling round his garden, determined not to give in to his infirmity. He had strong religious views, and brought up his children on nonconformist principles which they taught with undiminished vigour in their own families. The spirit, if not the form, of his fine puritanical creed lives in many of his descendants to-day, and in none is it more apparent than it was in John Hopkinson. It was the expression of a sturdy independence of character, and it taught an almost stern devotion to truth and duty. These were the qualities which Alice Hopkinson learnt from her father and passed on to her son. Of the narrowness which sometimes marred the nonconformist faith in those days there was none in her teaching. Thus it was, that though John Hopkinson in part rejected the form of her creed, he was able to keep its essence, and this slight divergence never affected the sympathy between him and his mother. To the end of his life it remained, a perpetual joy to both of them.

John Hopkinson's father, now Alderman Hopkinson of Manchester, was born in Lancashire and began life as an apprentice in the firm of Wren and Bennett, manufacturing engineers. In this concern he became a partner, held that position until 1881, and then devoted himself exclusively to consulting work. He was interested in several coal and iron enterprises and has been for many years Chairman of the Carnforth Iron Works. He has made himself an honoured name in connection with the affairs of his native city; his membership of the Council dates from 1861 and he was Mayor in 1882. As an engineer, he was brought up in a school in which practical experience was regarded as the one essential to success: but though in the first place practical, he had a remarkable sympathy with scientific development. His training made him attach much importance to precedent; but if conservative, he was by no means conventional in his ideas. He had much of the independence which so distinguished his eldest son. Especially was this the case in matters relating to the education of his children. In sending John to Cambridge while intending him to be an engineer, he went contrary to all the traditions as to technical training in which he had been brought up.

It is inevitable that in the first few years of childhood the mother's influence should be the more potent; and this was doubtless the case with John Hopkinson, though less so than usual: Mr Hopkinson, more than most fathers, made companions of his children, and John, the eldest, was constantly with him from a very early age. He was a teacher of exceptional ability, and during the long rambles in the Lake District and Wales and the visits to the Works when he accompanied his father, John must have learnt something of his powers of observation and have acquired some of his bent towards engineering and scientific pursuits. A singularly complete education rather effaced the surface effects of these early years, and it is perhaps difficult to find indications of their importance for those who have only known John Hopkinson as a man and have never known his parents. Others more fortunate, who have seen Mr Hopkinson with children even in his old age, will be able to form some idea of his relations with John at a time when the father was full of youthful vigour and the son a bright and healthy boy, eager for information on all going on around him. Though the first-born was naturally the most favoured one, all the family shared in the advantage of having such a father. Of the five sons, three, including John, became engineers of distinction. A fourth, Albert, is a doctor of medicine in large practice in Manchester. Alfred, who was second in order of age, became Principal of Owens College in 1898 after a successful career at the Bar and in Parliament. Such diverse successes in one family must be largely due to home influences; to the father is to be ascribed the aptitude for scientific and practical pursuits which has been so apparent in four of his sons and most of all in the eldest.

After a short time at a day-school near his home, John was sent to Lindow Grove in Cheshire. This school was then in the hands of a Mr Satterthwaite; but, in the summer of 1861, it passed into those of Charles Willmore, who, for some years after this, had charge of John's education and undoubtedly

exercised great influence upon him. His views on the training of boys were remarkable and are worth quoting here, as they furnish the key to much in John Hopkinson's early training. In a letter written to him in December 1866, Willmore says:—
"I greatly object to long hours of study for boys. I have a great belief in the physical development being the main thing till the time a boy leaves school; that is—for what I have just said might be misinterpreted—that any intellectual advance at the expense of health is very dearly bought." These doctrines he practised consistently without much regard for what the world thought of them. The school was situated in fine country and the boys were encouraged to spend much of their time out of doors. Thus the taste for open-air pursuits which Mr Hopkinson had given to John was further strengthened by his school-life, and this was beyond question the most important feature of that part of his education: its effects never left him. Throughout his life he rejoiced in exercise and fresh air beyond all things. As he himself put it, Willmore taught him the "self-regarding virtues," and that, in his case, put all the other teaching in the shade, good though it was. But for his healthy tastes, his rather delicate body could hardly have carried "that most fiery spirit." Of course he might have acquired them at one of the Public Schools—though there were not so many of them in country districts then as there are now. But it must be remembered that John Hopkinson did not belong to the class from which Public School boys are drawn. His father intended him to be an engineer; he had neither the inclination nor the means to let his son spend five years in enjoying himself and learning Latin verse at Harrow or Rugby. It was necessary that he should get as much knowledge of a useful kind as possible in the short time available before taking to business. Willmore was almost alone among schoolmasters in seeing that, even for boys of such a class, health and healthy tastes were more necessary still; and in acting on his belief. Unfortunately, parents did not all show such wisdom as Mr Hopkinson did, and realise the soundness of these views. Willmore's firm stand against all manner of cramming and excessive study for

examinations, perhaps cost him many pupils; but he never wavered from it and it has set an honourable and distinctive stamp upon his work.

So it happened that John Hopkinson had a thoroughly good time at school. Without being precocious, he had more than common intelligence and was able, without great difficulty, to hold his own with his schoolfellows at their lessons. The atmosphere of his home gave him a tendency to hard work; but he was by no means a book-worm, and had much time and energy to spare for enjoying life. One letter, written from Lindow Grove, may perhaps be quoted here: it gives an idea of what he was like—shows that he already had a sturdy, independent character:—"There was a row yesterday about jumping over the farmer's bank; after all, though, we have paid very cheap for the fun. The farmer, when he caught them, used all sorts of imprecations at them; he said if he ever caught them again he would take them to Macclesfield. I was not in the party that he met. I do not believe that he would have told if it had not been that when he shouted most of the boys of our party ran. I did not. I think if we had walked away quietly, he would have been satisfied, but as they ran he thought that we meant it to annoy him. We had a very great deal of fun, and we only got stopped going walks."

In 1864 Mr Willmore left Lindow Grove and became headmaster of Queenwood in Hampshire. John and some other boys accompanied him thither. The school was situated in wild and beautiful country on the confines of the New Forest. No better place could be found for carrying out Willmore's ideas of education. Unfortunately, discipline had been rather slack under his predecessor, and he had at first to be rather strict. John's letters were mournful for the first few weeks because he could not go for long walks in the surrounding country; he evidently felt this restriction deeply; but it was soon relaxed and then for a year his life was more enjoyable even than at Lindow Grove. Mr Willmore says that he developed greatly at this time both in body and in mind. He was great at bird-nesting and would go about it with much energy and daring, climbing the highest forest trees in search of eggs. Then in summer there were hornets' nests to be taken and butterflies to be caught. He used to tell a story of how he once caught some hornets for one of the masters. The insects were duly presented, shut up in a bottle: the master expressed a hope that "the cat's paw wasn't burnt." "No," said John, "and I hope the monkey's mouth won't be"; a retort of which he was always very proud. Similar touches of humour abound in his letters. He seems to have shown a decided bent for mathematics while at Queenwood. He was, in fact, rather beyond the teaching provided in this subject; though it was probably much better than could be found in most schools of the same class. He also received a fair grounding in Chemistry. In its teaching of Science, though it was rudimentary compared with what one finds now, Queenwood was almost unique at that time. Among its masters, it had counted Tyndall, Frankland, and Debus.

One may fittingly close this sketch of John's time with Willmore by giving his own views of it. They are recorded in a speech delivered at a dinner given by old Queenwood boys in honour of their schoolmaster. He says, referring in half-serious spirit to the virtues which they owed to Mr Willmore:—"I will confine myself to one virtue, yielding in importance to none, one which I suppose moralists would class as a self-regarding one, one which has not received a place in systems of ethics—I mean that of obtaining through life such enjoyment as it is capable of affording. As far as I can remember, Mr Willmore did not inculcate this virtue upon us, but I am bound to say that he gave us every opportunity of practising it, and he was never better pleased than when he saw us practising it with reasonable discretion.... I am sure that if it were possible for one of us to do a thing that was mean or shabby, one of the greatest deterrents to his being guilty of it would be to remember the burning scorn with which Mr Willmore would have regarded anything of the kind when he was at school."

Early in 1865, when he was $15\frac{1}{2}$ years of age, John Hopkinson began his student days at Owens College. At that time the education furnished by Owens was the nearest approach there was to what is now considered the proper training for an Engineer.

Cambridge had not yet awakened to her responsibilities in this respect and the more old-fashioned engineers thought that their sons should go into the shops at fourteen. Owens aimed at replacing the first three years of an engineer's "time" with a more liberal and wider education than could be found in the workshops. The spirit of the place was much the same as it is now, though the appliances for teaching were far less elaborate. If the great laboratories which are now the pride of the College were not yet made there was plenty of experimenting for lecture purposes within its walls. There (and at that time only there) the best of teaching in experimental science and in engineering subjects was combined with pure mathematics and other studies of an academic character. Sandeman, the purest of pure mathematicians, of whom it is said that Hopkinson was the only man who ever understood his lectures, Barker, more practical but still a typical senior wrangler, taught under the same roof with Roscoe and Schorlemmer the chemists and Clifton the physicist. These were the men who were Hopkinson's teachers just at the time when his studies were becoming important. He soon began to impress them with his ability. He showed up ahead of his fellow students in the College competitions from the first; even in German he got a certificate—evidence of general power, but not, I fear, of lasting proficiency. It was his mathematics which won his successes; he did not care for and could not remember the long string of facts, which then formed so large a part of chemistry; but he could do *sums*—a power rare among chemistry students, and therefore liberally rewarded with marks.

Hopkinson's examinations and his success therein were a somewhat remarkable feature of his life. They began with the Dalton scholarships in 1867, in which year also he matriculated at London University (coming out ninth in the Honours Division), and continued almost without intermission until he took his fellowship at Trinity in 1871. His first attempt was for a small exhibition at Owens in 1865 and it failed because he worked too hard in preparing for it. This little disaster, taken very seriously at the time, gave him a horror of cramming which never left him. In certain subjects, such as Botany, which were out of his line, but of which regulations required that he should show some knowledge, he would occasionally condescend to it, but in Mathematics and Physics he regarded it as quite useless even for the immediate purpose of impressing an examiner. Thus his academic successes were a real index of his remarkable quickness of mind and not, as is so often the case, merely evidence of a good memory and of great industry. He hardly ever missed being first in a competition in which he was seriously interested. Few men have made so large a sum as he did in this way: none have repaid it more fully to the world.

While at Owens, Hopkinson lived at home. His brothers were then, as they always remained, his principal companions; and he had few other intimate friends. John Hopkinson took a high view of his responsibilities as eldest son and at quite an early age he began to assist in teaching the younger members of the family. When they grew older they followed in his footsteps. Everything that he did, and especially new departures, such as going to Cambridge, had far-reaching effects as an example to his brothers. His brilliant success in all that he undertook gave them confidence in following him. Later still, when he came to occupy an influential position, he helped them in every possible way. He had very strong clan feeling, based no doubt partly on sentiment, but largely also upon knowledge of the power of a united and able family. In recent years this appeared even more in his relations with his own children. That which made him an ideal elder brother made him also a perfect father.

This is perhaps the place to say something of Hopkinson's life in the mountains, for it was there that his friendship with his brothers, and later with his children, found its most perfect expression. Reference has already been made to his tree-climbing proclivities while at school. During the vacations at Owens he began to find more magnificent game in the hills. The expeditions in North Wales and the Lake District were probably not so sensational as they afterwards became, but in the course of their long tramps the brothers must have come across many pieces of scrambling which would test steadiness of head and sureness of foot. Rock-climbing is probably the finest of all exercises in the variety of its demands upon bodily activity; and his apprenticeship in the art did much to consolidate Hopkinson's frame. He was of a very spare and wiry build with a long reach. When at the height of his powers few men were his equal upon difficult rock. It was not till comparatively late in life that he took to mountaineering as a sport, but when once he had tasted its joys he cared for no other recreation. His first visit to the Alps was in the Autumn of 1871 just after he left Cambridge, but the fascination of the snows apparently did not seize upon him then, and it was not until some years later that he began to spend his summer vacation there regularly. He reached the zenith of his strength and activity when he was about forty years of age. At that time he was a truly magnificent sight upon a mountain. I have seen him ploughing through soft snow up to his waist on the slopes of the Aletschhorn and tiring out men ten years his junior who had but to follow in his tracks. On other occasions he would lead, carrying the day's provisions, over rocks where his companions could scarcely follow him unloaded. He seemed to need no sleep, he would do two long ascents in consecutive days, and enjoy both thoroughly. He and his brothers soon got beyond the powers of the average Swiss guide. I remember several occasions in which we took one to make up our party, but it was always a failure; the poor man was too slow and generally had reason to regret his job, for my father was exceedingly impatient of being kept back. He took pleasure, but to no inordinate degree, in making new ascents; his sentiment in the matter was much the same as in scientific discovery. In the one case as in the other, he cared quite as much for his own feeling that he stood where no man had been before, as for public knowledge of that fact, and despised all squabbling for priority. The fascination of the sport for him perhaps needs no explanation; nothing else within his reach could seize his attention and require the exercise of all his bodily and mental faculties. If he rode on horseback, his active mind was always occupied with some point in his work. Golf, shooting, fishing had no attractions for him. Nothing seemed to serve him as a holiday and set him up for the next year's labours except a month in the

Alps. Though it was the immediate cause of his death I cannot regret that he was a mountaineer, for it was in the Alps that one came to know him best. There alone the enthusiasm and poetry in his nature would blaze forth unrestrained.

Mr Hopkinson intended that his eldest son should join him in his own works and to that end would probably have put him to practical work in the shops on his leaving Owens: he certainly had no thought of sending him to Cambridge. In those days a Cambridge education was considered to be waste of time for an engineer. The successful members of the profession had had no such education themselves; many of them had started as common workmen. The consequence was that most engineers, including John Hopkinson's father, did not even number College men among their acquaintance, much less thought of sending their sons to a University. The initiative in Hopkinson's case came from the boy himself. Young though he was, he had remarked that the engineers of the time suffered from the narrowness of their training. They began with practical experience, and if they kept pace at all with the developments in their Art following in the wake of science, they had to learn that science late in life. John determined that he would reverse the process and devote some time to the study of science before beginning to practise. Mr Willmore seems to have been the first to hint to him and to his father that he might score academic successes in a wider field than Owens. It was, however, principally due to Professor Barker that the idea of seeking at Cambridge his engineering education took definite shape in his mind. It was Barker's advice and encouragement which led him in the spring of 1867 to enter for a Mathematical Scholarship at Trinity College. He was first among some thirty competitors. His success settled any doubts which his father might have felt and in October of the same year he began his University life.

The Trinity Scholarship examination was the first in which Hopkinson met men other than his fellow-students at Owens. Among his competitors were of course some of the best of his year at Cambridge. His success in it therefore established his reputation for ability in a wider circle than Owens College, and he added to it by his performance in the First B.Sc. Examination, at London University. Then as now, this competition, owing to its high standard and the emoluments associated with it, attracted the best Cambridge mathematicians in their first and second years. Among those who were in the honours division about this time were Moulton and Harding—senior wranglers—and W. K. Clifford, who was second in the Tripos. Hopkinson was beaten for the Mathematical Exhibition by Pendlebury, who was a year senior to him at Cambridge and was subsequently senior wrangler, but he obtained sufficient marks to qualify for it and won the exhibition for Natural Philosophy and Chemistry. It is therefore not surprising that when he went to Cambridge he was already regarded as the probable senior wrangler of his year. For private tuition, he went to Routh, who had for some years held the first place as mathematical coach. His academic career was much like that of other men destined to be senior wrangler, though it was more than usually free from failures. He took in succession the Sheepshanks Exhibition, the Scholarship in Mathematics at London, and all the other honours and emoluments which fall to the lot of the first mathematician of his year at Cambridge. It is unnecessary to catalogue them here; they can be found in the University Calendars of the time. Perhaps, however, the Whitworth Scholarship deserves more particular mention. These scholarships were founded with a view " to the promotion of engineering and mechanical industry in this country." The examination was largely of a practical character and Hopkinson's success in it is a reminder that in his academic surroundings he did not forget what he had learnt in his father's works. Indeed he was in the habit of spending a large part of his vacations there, an excellent corrective to the Cambridge of those days with its horror of all things technical.

To one who is not specially interested in his Engineering work, Hopkinson's time at Cambridge and a few years after is probably the most interesting part of his life. At school and at Owens he was but a clever boy with few distinctive features except such as appear in his class-lists. Having regard to what he became, the process of his education is of interest; but his personality is undefined. On the other hand, from a few years after leaving Cambridge the uniformity of his success and the freedom from strife make his life of little interest to any but his intimates. But at Cambridge it is different: there is constant development, there is sometimes trouble and difficulty, which, if it appears slight to older men, was of vast importance to the youth who suffered it and no doubt had as great an effect upon him as the sharpest struggle with adverse fate on one more hardened. His letters from Cambridge, too, reveal far more of his inner self than at any other time of his life, and with their aid it is possible to form a fairly vivid picture of what he was like.

His work filled the largest place in his life. It has been stated that he came up to Cambridge with a considerable reputation. One of his contemporaries (Rev. Arnold Thomas) remembers his being pointed out in October of 1867 as the prospective senior wrangler, "a keen, strong face, without however any of the peculiarities of expression which are usually supposed to characterise the child of genius." To some extent he shared the popular belief in himself. In his first term he writes to his father:—"As for competition, there is plenty in my own year; though I am probably the best man now, I don't think I shall be able to hold that position long, much less three years and a half." A rare faculty of his, and one very apparent even in these days, was an unbiassed judgment as to his own powers. Not only did he know what he could do, but he did not affect ignorance. It is interesting to read his letters from Cambridge and to see how, as data accumulate, showing him to be the best man of his year, he coldly draws his conclusions and frankly avows them. It was then the custom, as it is now, for the coach to set his pupils a weekly paper of problems to be done at leisure. For this marks were given, and all the best undergraduate mathematicians of the year met in these competitions. Hopkinson faithfully recorded the results; during his first year he was often second and even third among his contemporaries. But it troubled him not at all; he still went steadily on his way with a single eye to the greater competitions to come. The fierce excitement of the Tripos perhaps affected him a little in the usual way. In the midst of it he writes:—"There is no doubt how the matter stands, viz.:—that if I had done up to my usual form, my place would have been safe, but as it is it is very likely I have lost it, and if one man beats me, two or three may." However, if he felt fear in that supreme moment he took no panic measures. "On the Monday evening," he writes, "for some reason—certainly not funk, for I have been cool as a cucumber the whole time—I did not get to sleep till about five in the morning, and as I was without chloral or spirits, I had no means of forcing it. However, I did not feel it on Tuesday. Last night I got some chloral and secured immediate sleep; having had one's brain in vigorous activity, it is rather hard to get the blood out of it." Even in such a crisis, he could regard himself as a physiological problem.

The result, in spite of his gloomy forebodings, was of course never in doubt. He turned out to have been first in fourteen out of sixteen papers. In the Smith's Prize he was again first. This examination was contrived to test originality of thought rather than the great speed which led to success in the Tripos. The papers were hard and the time for doing them was unlimited. Hopkinson's success in it therefore established his position beyond doubt as the first mathematician of his year. The year was thought by the examiners to be a fairly good one and judging from the subsequent careers of the men it was well up to the average, if not in the very first rank. That so eminent a mathematician as Dr Glaisher was second seems to imply that Hopkinson was much above the average of senior wranglers. His strong point of course lay in the physical application of mathematics. But the testimony of so high an authority as Professor Cayley, who was greatly impressed by Hopkinson's work in the Smith's Prize examination, shows that he was little inferior in pure mathematics. Hopkinson himself took little interest in that branch of the science; he used to compare it irreverently to chess-problems, the elegance of whose solution depends chiefly on the smallness of the number of moves. He had no appreciation of any originality or beauty in the problem itself. Still he looked upon the whole thing from a business-like point of view, and was not deterred from thorough study of a subject by any doubts as to its ultimate value. His principal aim at Cambridge was *to come out top*. His subsequent career is a triumphant vindication of competitive examination, for surely few men have been so completely under its influence as he was for five years of his life.

This concentration of his energies on the one aim was one of several causes which prevented him from making many friends at Cambridge. He was one of a large family, with brothers near his own age, and there was therefore no great inducement to him to seek society away from home: his inclination was quite the other way. Probably too, on their side, his fellow-men did not fully understand him. They admired the tremendous energy with which he would row a hard race on the river and then, on the same day, go in and win a scholarship; but it was wonder, rather than sympathy, that they felt. One can well imagine too that in those days he would be at no pains to conceal his impatience when he came into contact with less vigorous minds. It was in no sense "side"—the greatest of all sins in undergraduate eyes—it was that he was driven forward faster than others could follow and it was positive pain to him to hold back. He writes in 1869:—"The meeting of the Ray Club I went to was awfully slow. I had some talk with Prof. Babington at the beginning of the evening which was interesting. I then did nothing but gaze about me for half-an-hour, and finished up by sitting down to a book till it was time to go...." His was not the easy-going temperament which is the right medium for College friendships. Later in the same letter he remarks:—"I have lots of work on now and plenty to think of, what with Mathematics, boating and other engagements; I like a certain amount of pressure," and he frequently says that the place would be intolerable without plenty to do. It is easy to understand that such a character would provoke the respect of undergraduates, but not their love.

That he stood somewhat apart from the majority of his fellows did not affect him greatly. Once in a moment of trouble he alludes to it and in a rare fit of introspection seeks for its cause in himself. But that one allusion shows that he felt the need of sympathy in those around him, and though he cared nothing for general popularity he was by no means a recluse. Nor was the pursuit of his primary purpose at all inconsistent with the fullest enjoyment of every side of life at Cambridge. He made several intimate friends among the mathematicians, notably J. W. L. Glaisher, and Carver, who was fourth in the Tripos. Towards the end of his time, when his great powers commanded attention, he came to know such men as James Stuart, J. F. Moulton, W. K. Clifford, and he was made a member of the famous Apostles' Club. He was a good oarsman, and became captain of the Second Trinity Boat Club in 1869. On the running-path he won several races,—he was especially proud of a mile won shortly before his Tripos. His description of a walking race, in which he owed his success to his tactics as much as to his speed, is worth quoting:—"At the end of the first round—each round is one-third of a mile—Marriott led by about 10 yards. I did not see the fun of that, so I caught him and passed him as rapidly as I could, then he kept sprinting to catch me, to encourage which, I slackened at each corner and let him try to pass me. So for about eight corners he tried to pass me and walked on the outside in consequence. I found this manœuvre very useful: for the last two rounds he was content to walk just behind. About 150 yards from the end I put on the spurt and left him some 10 or 15 yards behind."

He had his full share of the high spirits which are supposed to be the glory of healthy British youth; and many were the escapades into which they led him. In one of them—perhaps remembered by his companion, who is now one of the great ones of the College,—"sulphuretted hydrogen, a horribly bad-smelling gas," was generated in the next man's rooms, wherefore porters and plumbers came to take up the drains, and the University Professor of Latin was driven from his quarters. His conflicts with the authorities were many and various; though, as a rule, they were wisely condoned. Against compulsory chapels, which were then more of an incubus than they are now, he waged unceasing war. To the end of his days he carefully preserved his correspondence with the Dean; who appears, however, to have been a person of great discretion, and not much disposed to exact the letter of the law from a prospective Senior Wrangler. Hopkinson's attitude in such matters was no doubt largely that of the healthy undergraduate : but it was also based on a mental characteristic, which outlasted youth. To the end of his life he had a hatred of all conventions, unless one must so name the moral principles which were part of himself and of which he was hardly conscious. Any law or custom whose breach would not be inconsistent with these must be justified to his reason, or he would thrust it aside not only without compunction, but with keen pleasure. He took a simple delight in mildly shocking people, as all who have seen him in his lighter moments will remember. At Cambridge he greatly relished his position as a Nonconformist and Radical among the Tory Churchmen, and seized every opportunity of playing the Philistine. The same spirit often appears in a more serious form in his manner of dealing with scientific questions. It was not only that he was original; he took a positive pleasure in working away from traditional lines, because they were traditional. There is no doubt, for example, that he found additional zest in his magnetic researches, because they showed the elaborate mathematical hypotheses of the subject to be without any basis of fact. If he could ever be said to be affected by prejudice in such matters it was against accepted theories.

Hopkinson soon made up his mind that it would not suit him to spend his days as a Cambridge don; and in his last year he began to look about for a profession. He had some thoughts of taking a Mathematical Lectureship at Owens College, which was vacant about this time. The letter which he wrote to Principal Greenwood asking for information about the place, and stating his qualifications for it, is of much interest:—" My qualifications for the post are a thorough knowledge of pure and applied mathematics so far as examination can test such a knowledge; together with a better knowledge of science generally, especially general Physics and Chemistry, than is possessed by most men of equal Mathematical attainments. On the other hand I am not accustomed to the use of instruments of physical research, have had no experience of teaching, and am much younger than is usually thought suitable for such positions. But all these objections are less serious when the head of the department is a man of such high standing as Professor Balfour Stewart, and the last is thus far an advantage that I shall have had no time to learn to be idle. Unless it is intended immediately to very greatly extend this department of the College, I imagine there will not be enough work completely to occupy the time of the second professor. This would give me two advantages in my candidature: first, if another man had the post it would probably be necessary to bring him to Manchester to do work which would not sufficiently occupy him; second, what surplus time I may have on leaving Cambridge, I intend to devote to acquiring such knowledge of engineering as will give a more practical turn to the studies I have been engaged in here, the being partially engaged in such work would, it seems to me, be valuable to a professor in Owens College." His principal motive in applying for the post was probably a desire for independence of means; he certainly thought that his career was to be found in engineering and did not regard this professorship as more than a temporary expedient. His father, however, with rare breadth of view, saw that even so it would unduly cramp his energies and advised him to give it up. As Dr Routh was of the same opinion he withdrew his application. His election to a Fellowship at Trinity in the autumn of 1871 gave him the independence he wanted; though the condition of celibacy, which was then attached to it, was destined to limit his tenure to some eighteen months only. It is interesting to note that the fellowships of 1871 were the first awarded without the imposition of a religious test.

On leaving Cambridge Hopkinson at once proceeded to carry out his own and his father's intention by starting work in Wren and Hopkinson's factory. It was not long, however, before another and better opening appeared. Many years before, the firm had made glass-grinding machinery for Messrs Chance Brothers and Co., Light-house Engineers, of Birmingham, to

Mr Hopkinson's design. Mr James Chance, the head of the firm, was himself a mathematician of much ability and had a distinguished academic career. From time to time he heard with interest of John's successes at Cambridge and when, in 1871, he wished to be relieved of the active superintendence of the Lighthouse Department, his thoughts turned to young Hopkinson as a likely man to take his place. He commissioned Mr Hopkinson to make an arrangement with John under which the latter was to become Engineer and Manager of the Optical Works for a year. It was arranged that he should spend a few months at his father's works in order to acquire some practical experience; and at the end of that time, in March 1872, he began his professional life in a position of much responsibility. It was no light undertaking that confronted him. He was not yet 23 years old and had had little engineering experience. He was put at the head of men many of whom had spent a much longer life than his at the work in which he was to supervise and command them. Yet within three years of his appointment he was not only acknowledged master, but he had largely increased the profits of his department by various changes in the methods of management. The rapidity with which he gained both the respect of his subordinates—from whom, as he said, he had everything to learn—and the confidence of his employers, was extraordinary. It was doubtless due in part to the fact that his position was one peculiarly fitted to his powers—in no other branch of engineering could one so favourable have been found. If, for example, he had begun in railway engineering, he would have had to spend much time in learning the accumulated experience on which the art is principally founded, and without which no man, whatever his genius, can attempt to practise it. Though his ability and energy must have brought ultimate success, it could hardly have come so quickly as it did. But the design of lighthouse apparatus rests, in the first place, upon mathematical deductions from experiment on the optical properties of glass. Experience, if useful, is not so essential for its accomplishment. To a man of Hopkinson's attainments the mathematics were a simple matter;

his training at Owens directly fitted him for the experiment. Thus in one important part of his work he had little to learn. It was of course by no means all, but his preeminence in it at once gave him a status with those above and below him and more than atoned for his lack of experience in other departments. Nor was the latter want by any means an unmixed evil. To one of his vigour of mind and character six months sufficed to supply it to a great extent; and then he was none the worse for having come to his work fresh and unhampered by tradition.

During the year 1872 he lived alone at an inn a few miles from the Glass-works. There is, in his correspondence for that year, an almost daily record of his doings. He was at first put in charge of the Lighthouse and Optical Glass Departments: the Coloured Glass Department was subsequently added. In addition to designing the lighthouse apparatus, he had to superintend its manufacture from beginning to end. Glass-making, the grinding and polishing of lenses, the design of iron towers—all these matters came under his management. In no one of them had he previous experience; but by the end of that first year not only did he know them soundly, but he had left his mark on each. As early as July 1872 he began experiments with the intention of improving the quality of the optical glass, which, as then manufactured, was of poor colour. Six months later he writes that their leading customer in this line is in high good humour with the glass they are supplying. So it was throughout his work. Even in this first year, when most men would have been content with learning the job and keeping the place going, he introduced change and improvement in every direction.

There is something very attractive in the picture of a young man in a position of great responsibility, presented in that series of letters. The responsibility one has to infer—it sits so lightly on him that it is rarely mentioned. But the youthfulness is very evident. Every new problem is hailed with delight and attacked without a trace of fear or anxiety. One sees, so to speak, only the overflow, and is left to imagine the further depths of energy and power necessary for the routine work. The technical questions which came before him interested him most. The management of the men was less to his taste, and he found the settlement of wages rather a worry. Yet even here he inaugurated a new order of things, introducing a system of premiums on the work turned out in less than a specified time; and in due course he got piece-work in operation throughout his department. This was in some ways his most remarkable achievement in this year, as it was certainly the most valuable in its results both to the men and their employers. In many parts of his work he found his father's ripe experience of great assistance. During this year he constantly asked for Mr Hopkinson's advice, both on technical points and on his relations with his employers. He always had great confidence in his father's judgment in such matters, and to the end of his life would often refer to him for help.

The lonely year at Birmingham ended in March, 1873, when Hopkinson married Evelyn Oldenbourg, to whom he had become engaged a year previously. He had now a wife and home, a considerable income and an established position; and this at a time of life when most young men are occupied in struggling for one or more of these desiderata. He was free from all anxiety for the future. He had satisfied sufficient of his ambitions to give him confidence that the others would be in due course satisfied also; and meanwhile he had the joy which attends striving when success is almost certain. Such a frame of mind was the best possible for his work, and to it in no small measure must be ascribed his early productiveness. How far that extraordinary power of concentration which was the secret of his rapidity of work would have been proof against constant personal worries, how far it was due to their absence it is difficult to say. It would seem that freedom from anxiety was essential to his scientific work at any rate: at the time when he was contemplating removing to London for instance, he remarks in a letter:—"At present being unsettled I am doing nothing in science," and similar expressions occur on other occasions.

Hopkinson remained with Chance Bros. until the end of 1877.

Swan Electric Engraving Co.

During that time he designed and manufactured the optical parts of many lighthouses which are now working in various parts of the world*. If one had to point to concrete objects as evidence of his power one would choose these lights. They do not perhaps appeal to the imagination as does a great bridge, but there is about them a certain finality and completeness which makes them no mean monument. The chief feature which Hopkinson originated in lighthouse practice was the group-flash system. This invention like many others of great practical value was perhaps less remarkable for its ingenuity than for the exact appreciation of what was wanted which prompted it. The various crude methods of identifying lights which preceded it are described in Hopkinson's pamphlet on the subject†. In 1872 Sir W. Thomson read a paper at the British Association in which he proposed as an improvement the use of long and short flashes according to the Morse code. In the discussion which followed Hopkinson vigorously criticised this proposal. It does not appear from his account of the incident whether he then advocated the group flashing, but it must have come into his mind about this time. Whether the idea was wholly original is not quite certain; there is no doubt that some experiments were tried in Belfast, in which a gas-light was turned up and down rapidly during the revolution of an ordinary revolving apparatus, thus producing a group-flash of somewhat uncertain character. These were published in 1872 as evidence of the value of gas in lighthouses and did not lead to any result except, perhaps, to give Hopkinson an idea. The merit of his achievement is untouched by them; it lay in his appreciation of the value of the idea, and the neat modification of the optical apparatus by which he put it into practice without great expense or loss of efficiency.

Many other points of detail in lighthouse construction such as friction rollers, clock governors and the like received substantial improvement in Hopkinson's hands. Some of them appear incidentally in his papers and it does not seem necessary to refer to them further here. His lighthouse work was remarkable

* Among the more important ones in the British Isles were those on the Casquets, Longships, St Catherine's Point (Isle of Wight) and Bull Point (Devonshire).
† Vol. I. p. 1.

relatively rather than absolutely. Very striking as coming from one so young and with so little experience, it would have been less so in a man of forty of equal mathematical attainments and brought up to works management and mechanical design. Good though it was, it had not the original value of the pure scientific work done at the same time to which we will now turn.

While at Cambridge and in the first year at Birmingham Hopkinson had published several pieces of original Mathematics. They are interesting now because they were his work rather than from any great intrinsic value. It is from this point of view that some of them are reproduced here. They exhibit a decided tendency to make the mathematics fit the facts, and therein they contain the germ of what became Hopkinson's most marked characteristic as a scientific man*. His mind had been influenced by two schools of thought. On the one hand there were the engineers, who distrusted theory and did not appreciate its value even as a means to practical ends: on the other was Cambridge, where the theory was considered the thing, while the facts to which it was to be applied were of minor importance. Until the application of electricity there was little sympathy between the two. A few great Cambridge men—Stokes, Thomson, Maxwell—had gone so far as to make mathematics a real instrument in physical research. But Hopkinson was the first to go further and to give both mathematics and experiment their true places in Engineering. A third school—Owens College —professed to combine the best features in the teaching of the other two and undoubtedly had much effect on him. But it needed the object-lesson furnished by his work at Birmingham to complete the union. That the elements of his education were still slightly discordant when he went to Birmingham is shown by his remark that he used to feel a sort of pleased surprise when he found that his calculations came out right and that his lenses really distributed the light as theory said they would. This feeling soon gave place to a most complete confidence in the power of mathematics and of laboratory experiment as engineering tools. At the same time he learnt their limitations; his contact

* See especially those on the breaking of an iron wire, vol. II. pp. 316 and 321.

with hard facts eradicated any lingering regard for hypothesis for its own sake which Cambridge may have planted in him. If his science strongly influenced his practical work, the practical work reacted no less strongly on the science. The long train of pure research on electrostatic capacity took its rise in experiments on the glass which he made at the Works; and similarly in later years his work on magnetism was preceded and prompted by enquiries into the properties of iron with a view to the construction of dynamo machines.

It was Sir George Stokes who inspired Hopkinson's first experiments on glass. They had a practical object—that of making glasses which should achromatise with one another without producing a secondary spectrum. It had been found that the dispersion of a phosphatic glass could be corrected in the required way—that is, the blue end of the spectrum could be extended without a proportionate increase in the total dispersion between blue and red—by an admixture of titanic acid. A perfectly achromatic telescope had in fact been constructed of glass prepared in this way; but its softness was a bar to its general use. Sir G. Stokes thought that similar effects might follow the addition of titanic acid to ordinary silicic glass; and made the suggestion to Hopkinson. The glass was duly made at Birmingham but the expectations were not realised*. For Hopkinson the most valuable result of the research was perhaps the long correspondence which it occasioned with one of the first scientific men of the age.

This work went on during the year 1874 and part of 1875. At this time Hopkinson saw much of Sir William Thomson, who more than any other man was to influence the direction of his thought. The occasion of their first meeting has been mentioned above—when the young man stood up and attacked Sir William's view on lighthouse illumination. Sir William was keenly interested in the subject; he had himself done much for mariners in other directions and was an enthusiastic yachtsman. Thus there was from the first a bond of sympathy between the two men, which widened rapidly as they became more intimate. Of Sir

* See Vol. II. pp. 341, 342.

William's kindness in these days and of the interest he took in his work Hopkinson always spoke with the warmest appreciation. He felt and expressed the greatest admiration for him. In one of his letters written after a visit to Glasgow he says: "It does one much good to come in contact with such a man; with most scientific men I meet I feel as though I had as clear a view into Natural Philosophy as they, but very far from it with Sir William Thomson." The force of such words from John Hopkinson can only be adequately appreciated by one who has read through his correspondence. Of the other men he met he said but little, but it is quite clear that Thomson was one of the few—two or three at most—whom he felt to be his equal in mental calibre.

Another potent influence in Hopkinson's intellectual life at this time was Maxwell's *Electricity and Magnetism*. The first edition of this work appeared in 1872. For a long time his evenings were regularly occupied with intense study of the book, and to it and to Sir William Thomson's personal influence is directly traceable his early work on Electrostatics. It is probable that from the first two leading ideas were present to him directing the course of his experiment on this subject. One is to some extent foreshadowed in one of the suggestive hints in which Maxwell's book abounds. Speaking of residual charge, "it may indicate a new kind of electric polarisation of which a homogeneous substance may be capable, and this in some cases may perhaps resemble electrochemical polarisation much more than dielectric polarisation*." Hopkinson went further and said that dielectric polarisation with its attendant residual charge, and electrolytic conduction, were perfectly continuous phenomena—so closely related that it was impossible to say where one began and the other left off. The other point which, though connected with the first, appears to have been entirely the creation of Hopkinson's own brain is that to residual charge is to be ascribed the difference between the capacity of a dielectric as ordinarily known and measured and its capacity as inferred from its refractive index according to the Electromagnetic Theory of Light. Knowing as we do that the capacity of glass varies with the time of electrifi-

* "Treatise on Electricity and Magnetism," Art. 331.

cation it is in no way remarkable, but rather just what we should expect, that glass should "deviate from Maxwell's law." Hopkinson's great generalisation, to which half his scientific work was directed, is that bodies in which such deviation occurs also exhibit residual charge and (under proper conditions) electrolytic conduction. He began by investigating the slower changes of capacity with time, he then carefully determined the effect of the more rapid changes in making the specific inductive capacity as ordinarily measured greater than the square of the refractive index, and finally, but a year before his death, he found a substance in which the whole process could be traced from beginning to end. He discovered that the capacity of ice which, after electrification for $\frac{1}{100}$ of a second, is of the order of 80, falls to about 3 if the time be $\frac{1}{10000}$ of a second. In other words, the great deviation from Maxwell's law observed in this case is almost wholly due to Residual Charge coming out in the times dealt with in ordinary electrical experiment.

The first paper on Residual Charge* described some qualitative experiments and formulated a working hypothesis as to the nature of the phenomenon. Sir William Thomson had long previously suggested that a dielectric under electric force might be regarded as in a polarised state similar to that of a magnet and so explained specific inductive capacity. Hopkinson supposed that a portion of this polarisation took time to develope, or to decay after removal of the electric force. The remanent polarisation he regarded as analogous to the coercive force of a magnet and this analogy led him to try the effect of mechanical agitation which in a magnet destroys the coercive force. He found that tapping a Leyden jar which has been charged, discharged, and then insulated, has the effect of accelerating the appearance of the residual charge —undoubtedly the most notable point in the paper. In the second paper†, published nine months later, the polarisation theory gave place to a much more comprehensive way of looking at the facts. In his book Maxwell treated conduction in dielectrics as made up of the displacement currents due to capacity and the conduction

* Vol. II. p. 1.
† Vol. II. p. 10.

currents which are observed after long application of electric force. This way of regarding it was inadequate, as it gave no account of residual charge except by making use of the artificial hypothesis of a stratified dielectric. Maxwell himself looked upon the latter as nothing more than a model and personally suggested to Hopkinson the mode of expression which he adopted in the second paper. Suppose a dielectric to have been continuously under the action of dielectric force, and that X was the magnitude of the force t seconds ago. Then the resulting electric displacement is properly expressed as $\int_0^\infty X . \phi(t) . dt$, where $\phi(t)$ is a function of t only, whose form depends on the nature of the dielectric. The most valuable part of the second paper is the experimental proof in the case of glass of the assumptions involved in this expression —that residual charge is proportional to the electric forces which produced it, and that the effects of these forces may be added if they were applied at different times.

This one expression includes capacity as ordinarily known, residual charge, and dielectric conductivity as ordinarily known. It constitutes therefore a statement of the first generalisation mentioned above—that these three things are really parts of one continuous phenomenon. That Hopkinson should adopt it is eminently characteristic of his scientific methods. He cared only to *express* observed facts and their relations; any hypothesis that went further than this appeared to him useless and often mischievous. Though consistent with what was known it might well turn out to be at variance with the unknown when that came to be discovered. Analogy and hypothesis he would use as a means of suggestion, but in so doing he took care never to lose sight of the foundation of his work—a mode of expression of the known, sufficiently wide to be consistent with the unknown, whatever it might turn out to be. He did not like the old way of looking at conduction because it admitted only two constants, capacity and conductivity, and had to invent stratification in order to explain Residual Charge. He preferred rather to *express* the latter phenomenon by the introduction of fresh dielectric constants; it seemed quite arbitrary to regard it as less fundamental than the capacity and conductivity on which it was sought to base it. That he was right is shown by the *reductio ad absurdum* in which resulted the attempts at applying the old theory to the facts of metallic reflection. Hopkinson's way of looking at the matter leads to results which are quite consistent with observation. In the same way, though he was convinced of the truth of the Electromagnetic Theory of Light, it was in his mind shorn of all reference to a medium. To him electric and magnetic forces were simply directed quantities in space, and light an electromagnetic phenomenon. It was as wrong to "explain" the electric forces as due to displacement and pressures in a material medium as it would be to explain the pressure between bodies as due to electricity. The two classes of phenomena were to his mind equally fundamental.

In the year 1876 Hopkinson published some very careful determinations of the refractive indices of various glasses*; and this was closely followed by an account of his first experiments on its Specific Inductive Capacity†. The latter paper opens with a statement of the relation between the two quantities predicted by the Electromagnetic Theory of Light and then proceeds to the description of observations, showing that in the case of glass μ^2 and K are very far from being equal. There is no comment on this result except a warning against the inference that Maxwell's Theory is (in its more *general* characters) disproved thereby. Hopkinson's work on Residual Charge and Capacity led to his election as a Fellow of the Royal Society in 1877.

In 1877 Hopkinson decided to remove to London. He did not sever his connection with Chance Bros., but made an arrangement with them compatible with residence away from Birmingham. This helped him considerably in what was otherwise a very bold step, for his family responsibilities had increased greatly in the five previous years, and he had little to depend upon except his brains and growing reputation. Work came very slowly at first in the new field, but when it once began it increased rapidly, and in the first six months of 1880 he did better than in the previous

* Vol. II. p. 44.
† Vol. II. p. 54.

two years. The most lucrative part of his practice was, as it always remained, expert evidence in patent suits. Hopkinson's powers were of a kind to ensure his success in this profession. In concentration and rapidity of thought he was of course pre-eminent. The fighting instinct in him made him delight in the exercise of these faculties when under cross-examination by a clever counsel. But perhaps his capacity of drawing the line between fact and argument was that which most distinguished him. Expert evidence is a curious mixture of these two elements, and the best experts are those who can separate them. Some men cannot avoid usurping the position of the advocate and take a side about what the facts *are*, thus discrediting themselves with the judge. Others fail to make the most of the controversial points as to the interpretation of the facts, which though supposed to be matters of fact, and of law for the judge, are really questions of argument in which the expert witness takes a large share. Hopkinson, alike in the witness-box and in the laboratory, knew exactly how far he was bound by the facts and where difference of opinion could come in.

The telephone cases were the first important ones in which he was engaged; the case of Attorney-General *v.* The National Telephone Co., in which it was decided that telephone messages were part of the Postmaster-General's monopoly had more than a passing interest. A portion of Hopkinson's evidence in that suit is worth quoting as a fairly representative example of his style. "If the telephone exchange is not used otherwise than is described in the affidavits, and as I am informed and believe, no message telephonic or other is transmitted by its agency or along the wires occupied by its subscribers. When A and B are conversing by the telephonic exchange they speak directly with each other. A message implies an agent who conveys the message. If A, having no speaking trumpet, asks C who has one to say something to B at a distance from him the communication may be called a message; but if A pays C a shilling for the loan of the instrument and speaks himself, it is no message. Replace the trumpet by the telephone and wires, and the same distinction holds. I therefore say that it is untrue that messages are transmitted by the telephonic exchange." Another case which will be remembered was Easterbrook v. Great Western Railway Co. The patent in question was for a system of interlocking railway signals. Hopkinson, who acted for the Defendants, discovered that it was possible by certain movements of the levers to put the signals at variance with the points. The Plaintiffs, wholly unsuspecting, had closed their case; and their consternation may be imagined when Sir Frederick Bramwell expounded on their own model how their system would lead to a collision. The patent was of course upset. Other achievements less dramatic but as useful to clients soon gave Hopkinson a wide reputation, and for many years he was employed in every case of importance.

Some may be inclined to regret that Hopkinson was so much employed as a patent expert. This is not the place to discuss the usefulness or otherwise of that kind of work; but it is at all events desirable that it should be done well, and Hopkinson set a standard of well-doing that will not soon be forgotten. Apart altogether, however, from the possible intrinsic worth of expert work done as he did it, it had through him an indirect value which fully justified his adoption of it as part of his profession. He did it with extraordinary ease and rapidity, and if it had no lasting results its comparative freedom from responsibility and the substantial income derived from it enabled him to attack his engineering and scientific problems in peace of mind, undisturbed alike by professional worries and family responsibility. I think that this absence of anxiety was of great importance to him. Had he depended for his living upon constructive work of a more wearing character its results might have been more useful than what he did in the Courts, but it is doubtful whether he would have risen to such heights in other directions. His patent work was of considerable use too in keeping him in touch with a great variety of engineering and manufacturing practice. Experience so acquired is of course of a very superficial character and of little value to one who has not commanding originality, but to Hopkinson it was frequently a means of suggesting new advances, and it was a strong connecting-link between his scientific thought and the world of practice.

When Hopkinson came up to London in 1878 Electric Lighting was just emerging from the laboratory and a very limited sphere of application and had hardly become a commercial question. Its practice had advanced little since 1851, when it was applied to Lighthouse illumination; its theory rested almost where Faraday had left it. The discoveries of Wilde, Siemens and Wheatstone as to the excitation of dynamo machines, important as they afterwards became, had as yet little practical significance. Furthermore these discoveries and indeed almost all that was known to engineers about electrical machinery were of a qualitative character, the quantities concerned were not understood and had only been measured in the roughest way. Still there was a vague impression among engineers that Electric Lighting had a great future which led them to give no little attention to it and to look eagerly for some means of applying it more generally. The invention of the incandescent lamp in the early eighties supplied that means and at once gave importance to everything concerning Electric Lighting. There naturally followed a demand for the accurate measurement and scientific treatment of the phenomena thus suddenly brought into prominence, and it fell to John Hopkinson to supply this want. One is struck by the fact that he, the man fitted above all others for the task by genius and education, came at exactly the right moment for its fulfilment. He had two or three years before the excitement of the electric lighting "boom" in which to quietly study the problems involved. When the "boom" came, bringing with it the stimulus of universal public interest and the rapid development due to the application of capital, he had done much of his work. He was perfectly prepared to attack the new questions which were constantly arising, and he was already recognised as one of the first authorities on the subject, so that everything he said and did took effect.

It is difficult at the present day to imagine the chaos which he reduced to order. The only quantities recognised or understood by the majority of engineers in connection with electrical machinery were the candle-power it could produce and the horse-power required to drive it. Early in 1878 a paper was read before the

Institution of Civil Engineers on dynamo machines*. In that paper reference is hardly ever made to an electrical quantity; where made it is vague and inaccurate. Current and potential were confused; it was gravely asserted—and Hopkinson was the only man in the discussion who contested the assertion—that the maximum theoretical efficiency of electric power transmission was 50 per cent. That paper gave rise to what was probably the first public discussion on Electric Lighting, and also I believe the first before the Institution in which Hopkinson took part. He had then no practical acquaintance with the subject, and he contented himself with formulating the questions which arose in his mind. What were the resistances of the armature and field coils of the dynamos tested? What was the power actually expended in the electric arc, and of the balance how much was wasted in the various parts of the machinery? These and such as these were the points, ignored by his fellow engineers, whose future importance he foresaw. To him belongs the distinction of pointing out what were the questions which must be answered no less than that of answering them.

It has been stated that one of the chief applications of the electric light in 1878 was to Lighthouse illumination. It was in that connection that Hopkinson made his first acquaintance with the art. Early in the year he began to enquire into the various forms of dynamo machines, and in the autumn he bought a medium sized Siemens dynamo for Chance Bros. Experiments on that machine were the subject of the first paper on "Electric Lighting," before the Mechanical Engineers†. There for the first time, logical expression was given to the properties of dynamo machines by means of the characteristic curve. The efficiency was roughly measured; it was shown how the capacity of a dynamo could properly be specified. A second paper, read before the same Institution twelve months later‡, was directed more particularly to measurements on the electric arc. On this point also there had been great confusion, and it will be found that

---

* "Some Recent Improvements in Dynamo-Electric Apparatus." Higgs and Brittle, *Proc. Inst. C. E.*, vol. LII. p. 36.

† Vol. I. p. 32.   ‡ Vol. I. p. 47.

both in the paper above referred to on Dynamo Machines and in a later one by Sir James Douglass on Lighthouse Illumination*, the brightness of the arc is almost always stated in terms of standard candles without reference to colour or direction. Hopkinson indicated the influence of these two factors. He also showed that the resistance of the arc was not constant, and pointed out several properties of dynamo machines which could be inferred from the characteristic curve. It may be said that in these two papers and in that on Alternating Currents read in 1884†, engineers were told how to find out all that it was useful to know about existing electrical machinery. The next step was to tell them how that machinery could be improved, and applied to new purposes.

In the early part of 1882 the newly formed English Edison Company appointed him to be one of their consulting engineers. The Company had been started by Americans with the object of working Mr Edison's and other inventions in this country. Prior to Hopkinson's appointment they had put down a small central station at Holborn Viaduct which was fitted with American plant. In those days the aims of the Company included the supply of all kinds of continuous current machinery as well as incandescent lamps; and they also took out several of the first Provisional Orders with the intention of starting supply stations, which however was never fulfilled. In his capacity as consulting engineer, therefore, Hopkinson had to deal with a great variety of electrical problems, and there is no doubt that his advice to the Directors of the Edison Company had a great effect in shaping the course of the electric lighting industry, though many of his prophecies were forgotten before they were actually realised. From time to time he wrote a report to the Directors on such scientific matters as concerned the business of the Company. These letters are very interesting reading; in them are set out the fundamental ideas underlying almost every point in the art of supplying electricity. It was long before many of his suggestions were appreciated, but they are now accepted as common-places. As early as June, 1882,

* *Proc. Inst. C. E.*, vol. LVII.
† Vol. I. p. 133.

Hopkinson stated the principles which should guide the undertaker in charging for current. In a report of that date to the Edison Company on the policy which they should adopt in relation to the Electric Lighting Bill he points out the great importance of the dead charges in central stations and urges the use of a double system of charge, viz. a fixed payment per annum dependent upon the maximum amount of current demanded by the consumer together with so much for every unit registered by the meter. This point he continually impressed upon his clients, and he secured powers to charge in this way in their early provisional orders. It was not, however, till some years later in the first stations which he himself designed, that it was actually put into practice. Now Hopkinson's method of charging with slight variations in detail is in use in half the stations in the country. Again, he clearly foresaw the impetus which the continuous current would receive on the introduction of a good storage battery, and he constantly urged the Edison Company to push the transmission of power, a great field open to them and at that time denied to their alternating current rivals.

The machinery first introduced by the Edison Company was of course manufactured in America. Hopkinson quickly perceived the defects in the early Edison machines and set to work to remedy them. Probably he had not then arrived at the idea of the composite magnetic circuit, but he had a strong opinion that length in the magnets beyond that required for getting on the windings was a disadvantage and that the machine should be designed so that the armature and magnets were magnetically saturated together. This idea must have been present to him when he wrote to the Company in September 1882:—" It is necessary to make a critical study of the (Edison) machines with a view not only of improving them but of placing ourselves in a position to say beforehand how we should modify a machine to meet varying conditions.......I am not without hope that we may succeed in largely increasing the output of the machines by alterations of the magnets." The matter was put to the test in his laboratory a few weeks later. Small models of the Edison machine were made to scale and also of the modified magnets and armature which

Hopkinson proposed. The inductions in the armature windings in the various forms were then compared by jerking the armature out of the magnet field and observing the kick on a ballistic galvanometer connected to its winding. It was found that in some of the Edison machines the armature and magnets were not saturated together, while much of the effective area of the armature was lost by bolt holes and the like, and that shorter magnets of the same cross section would give as large an induction with a smaller magnetising current. Hopkinson lost no time in urging the Edison Company to have a machine made with shorter magnets and an armature of the same weight but greater effective cross-section. This, the first dynamo constructed on scientific principles, was made by Messrs Mather and Platt, and the results completely justified Hopkinson's anticipations, for the out-put was more than double that of the Edison machine of equal weight and cost from which it had been evolved. In Hopkinson's report to the Edison Company of the tests upon it (May 17, 1883) one detects a note of triumph which he rarely permitted himself in such a communication:—"The new dynamo machine was last week tested at Messrs Mather and Platt's at full speed. It was found to be capable of supplying 200 ordinary 'A' lamps with ease when running at 1170 revolutions* and with an economy of power superior to any machine that has ever yet been made."

The Edison-Hopkinson dynamo and the statement of the principles which it embodied constituted a very notable advance; its greatest interest however lies in its origin: it sprang in full perfection from laboratory experiment; and it is a remarkable instance of the absolute confidence which John Hopkinson had come to have in such methods. Apart from the way it was arrived at it was of less importance; for it constituted but a small step towards the complete art of dynamo design. It stands in the same relation to the dynamo as did the discovery of condensing to the steam-engine but if one may continue the analogy there was as yet no application of thermodynamics—that was made in the great paper on dynamos two years later. Men soon began to cry out

* The Edison machine of the same weight and speed could only supply about half the number.

for more light. Their requirements were clearly formulated in the preface to the best known work of the time on dynamo machines*. "The true mathematical theory of the dynamo can indeed only be written when the true basis for writing it shall have been discovered. That basis, the exact law of induction in the electromagnet, does not yet exist...... Of the laws of induction of magnetism in circuits consisting partly of iron, partly of strata of air or copper wire, we know, in spite of the researches of Rowland, Stoletow, Strouhal, Ewing and Hughes very little indeed...... We want some new philosopher to do for the magnetic circuit what Dr Ohm did for the voltaic circuit fifty years ago." That is a clear statement of the difficulty which John Hopkinson and his brother Edward removed in 1886; Ohm's discovery is fairly analagous to their achievement.

It is probable that the law governing the induction in dynamo machines was in a general way appreciated by Hopkinson as early as 1882. In the notes of the experiments on the Edison machines above referred to there is a remark that a discrepancy between the behaviour of the model and the full sized machine is probably due to an error in the air-space of the former, which would of course be very likely to occur. It would appear therefore that Hopkinson then recognised the importance of the air-space term though he had not at the time any definite data as to its magnitude. Further notes at the end of 1883 and beginning of 1884 disclose all the essentials of the theory of the composite magnetic circuit, though the want of knowledge of the magnetisation curve of iron made accurate application of the theory impossible. The missing data were supplied by Hopkinson's experiments on the "Magnetisation of Iron," published in the *Royal Society Transactions* in 1885†. In that paper the fundamental equations of the composite magnetic circuit are given. The circuit used in the experiments contained two parts of different area and permeability, the specimen under examination and the wrought-iron block in which its ends were bedded. Given these equations it seems but a simple matter to proceed to the calculation of the

---

* "Dynamo-Electric Machinery," by Silvanus P. Thompson, 1st Edition, 1884.
† Vol. II. p. 154, and Appendix to Vol. I.

induction in a circuit containing air and iron instead of two kinds of iron. Perhaps, however, owing to the fact that the equations referred to form but a subsidiary part of the 1885 paper, and are only used for the purpose of investigatiug the correction due to the magnetic force in the wrought-iron yoke (a correction subsequently neglected), the solution of the further problem—the synthesis of the characteristic curve of a dynamo machine—came as a revelation to those interested in the subject. This was the ground-work of the joint paper in 1886*. Greater knowledge of the magnitudes of the various terms involved, however, revealed many discrepancies between the facts, and the inferences from the theory in its crude form. The careful investigation of these forms the greater part of the paper and involved much patient research. The spreading of the lines of force beyond the pole-pieces, waste field, and armature reaction were all exhaustively dealt with. Finally the amount and distribution of the losses in the machine were determined by the beautiful method of coupling two similar dynamos mechanically and electrically and measuring the power required to drive the combination. Not only did the Hopkinsons give the world a brilliant idea, but they instanced its application to every point of importance in dynamo design. Their paper is the complete and unassailable foundation on which that art has been built up. Almost everything that has been done on the dynamo since is in the nature of a rider on the propositions enunciated and proved by them.

It is a legal maxim that a principle cannot be patented, and the attitude of the world towards those who give it new ideas is much that of the law. It seldom happens that they reap so much material reward from their discoveries as those to whom they have taught them and who are thereby enabled to make new things. John Hopkinson however was to some extent an exception to the rule. He was able to take advantage of his own teaching and in his own person to turn it to practical use. His position was a favourable one, for his work on electric lighting and dynamo machines was the answer to riddles already existing for engineers and in some cases formulated by them. Having

* Vol. I. p. 84.

found the answer he could and did use it just as his fellow-engineers did after he had taught it to them. It was during the years 1880—1886 that he produced his most important electrical inventions. Though some of them became of great practical value, to his biographer they seem insignificant compared with the more general ideas produced at the same time to which they were mere corollaries. The most successful of all, the three-wire-system of distribution, was a happy thought which might have occurred to anyone who had Hopkinson's clear knowledge of electric lighting. The interesting point about it is the evidence it furnishes that he alone possessed that knowledge, not that possessing it he made the invention. The same remarks apply to the closed-circuit transformer and the series-parallel system of motor-control, either of which could hardly have cost him an hour's labour when once he had grasped the principles involved; though nowadays no transformer is made without using the one and no electric locomotive without the other. These discoveries however were not a pecuniary success, the transformer patent was of doubtful validity owing to a chance anticipation of ancient date, and the series parallel system was invented ten years before there were any electric locomotives. On the other hand the first electricity meter*, on which he spent a very large amount of time and labour, and which required far more ingenuity and power than the other three, was not a practical success. So it was with several other inventions into which he put a great deal of brains. He was in a sense too great a man to make a very successful inventor. Not only did he see too far ahead and so bring out things before the time was ripe for them, but he had not the sort of business instinct which correctly assesses the value of an invention, and makes the fortune of its possessor far more certainly than any amount of ingenuity. In his case this want resulted in a great expenditure of thought on things which yielded no return; while he reaped rich harvests where he least expected them. In the light of after events it seems remarkable to find the three-wire system patented along with an electricity meter, a means of

---

* Patents No. 49 of 1882, 3576 of 1882, and 3475 of 1883.

measuring potential and an automatic switch*; and the series parallel system of control a subsidiary claim in a patent primarily devoted to an electric hoist†. Neither of these epoch-making inventions was considered worthy of a patent to itself! Hopkinson used to say that on the whole he had been fairly paid for the time he spent on inventing; but the things which paid best were those which cost him least trouble.

Hopkinson's early association with the Edison Company, and his conviction that the continuous current was destined to fill a bigger place than alternating currents in future electrical distribution, made it inevitable that most of his practical work and inventing in the years 1880—1886 should be on direct current machinery. The theory of the magnetic circuit however, as well as other ideas which he originated or first brought to the notice of engineers in connection with their art, had no such limitations. Though he illustrated them at first by their application to direct current problems, they were equally applicable, and he applied them, to the rival system. Thus he was as much a pioneer in the theory of that system, if not in its practice, as he was in the domain of continuous currents. He was the first to show theoretically that two alternating current machines if run in multiple, would tend to keep in synchronism; thus bringing to light a fact discovered long previously by Wilde, but forgotten before it became practically important. This and many other properties of alternating currents were enunciated or proved in a paper read before the Institution of Civil Engineers, in April 1883‡. On that occasion as on many others, Hopkinson acted as interpreter between Faraday and Maxwell on the one hand, and electric lighting engineers on the other. It was a part for which his sympathy with both sides peculiarly fitted him. That sympathy constantly found expression. In a communication to the Royal Society, he would use analogies drawn from engineering—as when he suggested that the practice of telegraph cables should be followed in measuring the capacity of a flint-glass flask, the time

---

* Patent No. 3576 of 1882.
† Patent No. 2989 of 1881.
‡ Vol. I. p. 57.

of electrification however being one hundred millionth of a second instead of one minute*. On the other hand in the 1883 paper just referred to, which is addressed to Engineers, he begins with a statement of the laws of electric induction discovered by Faraday, which are beautifully illustrated by mechanical models, he goes on to apply them to practical questions, and the paper concludes with a comparison of the cost of lighting by electricity and gas. A further paper read before the Electrical Engineers in Nov. 1884, dealt with the mathematical treatment of alternating current circuits†. The problems solved were simple applications of the laws of induction, the machine being regarded as a circuit of constant self-induction moving in a simple harmonic field; but they were introduced to Engineers for the first time, and it was long before the solutions were improved upon. The more important and original work of deducing the properties of alternating current dynamos and transformers from their dimensions and the properties of the iron was done in 1887, a year after the same problem had been solved for continuous current machines. Like the improved dynamo, Hopkinson invented the closed circuit transformer some years before he fully investigated its properties in fact it was not until 1892 that he finished with it. In that year he made the first accurate determination of the amount and distribution of the losses in a transformer; using a method similar to that which he had devised in 1886 for the dynamo‡. His experimental work on the alternating current dynamo, too, came at a comparatively late period. The first good measurements of efficiency on such machines were some tests of his in Newcastle in 1895. Similar tests were made on a pair of Siemens machines in 1896, and at the same time he investigated the effect of the currents in the magnet poles induced by the armature currents§.

We must now return to John Hopkinson's purely scientific research, which never ceased during the years of which we have

* "On the Capacity and Residual Charge of Dielectrics as affected by temperature and time." Vol. II. p. 122.
† Vol. I. p. 133.
‡ Vol. I. p. 192.
§ Vol. I. p. 156.

been speaking. Down to 1885 his experimental work dealt mainly with electrostatic capacity. Further determinations of the capacity of glass, confirming his earlier measurements, were published in 1880*; together with the results of experiments on various liquids. As regards the latter he reached the general conclusion that the hydrocarbons agreed with Maxwell's law, while oils containing oxygen exhibited considerable deviation from it. He found in this result a confirmation of his theory that the deviation was due to something which might be indifferently described as residual charge or incipient electrolysis; for one would expect bodies like castor oil to behave like electrolytes, while paraffin would exhibit no such properties. A very interesting piece of work on electrostatics was a paper written for the Royal Society Proceedings in 1886*. Professor Quincke had made some determinations of the capacity of liquids by measuring the change of pressure of an air-bubble in a fluid under electrostatic force, and by weighing the attraction between the plates of a fluid condenser. He found a considerable difference between the values of the capacity determined by these two methods. Hopkinson examined this result, and showed that it could be explained either by assuming that capacity depended, like the permeability of iron, on the force, or by supposing that the capacity of a condenser did not vary inversely as the distance between the plates. The first assumption was negatived by some experiments of his own, the second was "subversive of all accepted ideas of electrostatics." He concluded by suggesting that the discrepancy was perhaps due to neglecting the capacity of the wires and key used in the experiments, a source of error which he had himself experienced. Correspondence with Quincke ensued, and that gentleman ascertained the capacity of his connections and found it to be responsible for the supposed discrepancy. Hopkinson was always very proud of this achievement, and indeed it was a remarkable piece of scientific criticism. One other paper should be noticed before we return to his work on magnetism, because it was so characteristic of his mode of thought. A controversy had long raged—indeed it is not yet set at rest—as to

* Vol. II. p. 65.   † Vol. II. p. 104.

the true nature of contact electricity and the seat of the electromotive force in a Voltaic cell. One party maintained that the difference of potential between different metals in contact should be measured by the Peltier effect in a thermoelectric couple. The other, led by Sir W. Thomson, held that it must be deduced from electrostatic experiments. John Hopkinson took neither side, saying that the question was "one of the relative simplicity of certain hypotheses, and definitions used to represent admitted facts*." One feels, in reading his development of this thesis, how utterly unbiassed he was in his scientific judgment. It was almost like the strife in the last century over the proper method of measuring the force of bodies in motion. "Shortly all Europe was divided between the rival theories. Germany took part with Leibnitz and Bernoulli; while England, true to the old measure, combated their arguments with great success. France was divided, an illustrious lady, the Marquise du Chatelet being first a warm supporter and then an opponent of Leibnitzian opinions. ......But what was most strange in this great dispute was that the same problem, solved by geometers of opposite opinions, had the same solution. However the force was measured, whether by the first or second power of the velocity, the result was the same†." Like d'Alembert in that historic controversy, Hopkinson showed that the whole dispute was a mere question of words.

Hopkinson's work on Magnetism began with the Royal Society paper in 1885 referred to above‡. The experiments described were intended primarily for practical ends, and found their first application in the theory of the dynamo machine. There were however one or two points of much scientific interest, notably the absence of the magnetic property in steel containing 12 per cent. of manganese. The fact was known to several people, among them Mr Groves of Bolsover Street, the instrument-maker, who brought it to Hopkinson's notice. Hopkinson made accurate determinations of the properties of the steel, and showed that the want of magnetism could not be ascribed to a mere admixture

---

* Vol. II. p. 375.
† "Elementary Rigid Dynamics," by E. J. Routh, 5th Edition, p. 277.
‡ Vol. II. p. 154.

of a non-magnetic material with the iron, even if the two substances were arranged in the manner most favourable to that hypothesis. The rest of the paper consists of determinations in absolute measure of the magnetisation of various samples of iron. The full value of this work will only be appreciated when some general law is found covering the isolated phenomena of the magnetisation of iron. In the present state of knowledge the determinations are of practical rather than scientific value, though their accuracy and the completeness with which the mechanical state and chemical composition of each sample is known, as well as the fact that they are in absolute measure, distinguish them from most other results of the kind.

Hopkinson's next experiments on magnetism, however, constitute a step, albeit a small one, towards the generalisation, the search for which seems likely to baffle men for a long time to come. It was known that the magnetic property of iron disappeared suddenly at about red heat. It was also known that if iron be heated to redness and allowed to cool, there is at a certain temperature a considerable liberation of heat not unlike, and comparable in amount with, that which occurs when water freezes. Hopkinson carefully examined these two phenomena and found that the magnetism disappeared with great suddenness at the temperature of "recalescence" as the liberation of heat is called*. Further experiments showed that both changes were accompanied by a sudden decrease in the temperature rate of variation of the specific resistance of the iron†. The true significance of this discovery cannot be better put than in Hopkinson's own words. "When iron passes from the magnetic to the non-magnetic state it experiences a change of state of comparable importance with the change from the solid to the liquid state, and a large quantity of heat is absorbed in the change. There is then no need to suppose chemical change, the great physical fact accompanying the absorption of heat, is the disappearance of the capacity for magnetisation‡."

In the course of these experiments on the influence of tempe-

---

* Vol. II. p. 217.   † Vol. II. p. 214.
‡ Address to the Institution of Electrical Engineers, Vol. I. p. 233.

rature on magnetisation, Hopkinson hit upon that extraordinary body, steel containing 25 per cent. of nickel. He found that this substance, which is substantially non-magnetic as it comes from the manufacturer, becomes magnetic if it be cooled below $-20°$ C., and remains so if after cooling it be allowed to return to the ordinary temperature. Its magnetic property cannot now be destroyed unless it be heated above $580°$ C. If that be done and it be allowed to cool again, it will be found to have returned to its original and non-magnetic state. Further investigation showed that many other properties of this steel were different according as it was in one or other of these two states. Its density, its elastic properties, and its electrical resistance could all be substantially changed by heating it to $580°$ C., and could then be restored by cooling to $-20°$ C. or *vice versa*. In fact almost the only property common to the two states was that of containing 75 per cent. of iron and 25 per cent. of nickel.

This discovery of a substance, non-magnetic as ordinarily known, which could be rendered magnetic by cooling seemed to open up an endless vista of possibilities. Why should not other metals exhibit the same property if treated in a similar way? John Hopkinson's imagination was kindled by that thought in a manner rare with him. I well remember the time when he put it to the test of experiment. Rings of chromium, non-magnetic manganese steel, and other metals, which from their chemical similarity to iron or other causes seemed worth trying, were put into solid carbonic acid and tested magnetically by the ballistic galvanometer. The results obtained were wholly negative; no one of the metals tried exhibited any sign of magnetisation. I think that on that occasion even Hopkinson came near to feeling disappointed with the course of Nature. The nickel steel with its twofold character, like iron with its magnetisation, remained in a state of inscrutable isolation.

Hopkinson's work as an inventor was practically confined to the decade 1879 to 1889. During those ten years his name appears in the patent records some five and twenty times, whereas he only took out six patents during the remainder of his life. His scientific work in the latter period shows that this was not due to any decay of original power. It was part of a general change in his professional position following necessarily upon the development of the electrical industry. In the early eighties Hopkinson was almost alone in his grasp of the correct principles underlying the new applications of electricity. That period was marked not only by much invention, but by papers to the Engineering Societies of a general character in which he gave to the world the principles which had guided him in inventing. As the field of electrical engineering widened, there was a corresponding increase in the number of qualified workers in it. They owed much of what they knew to Hopkinson; and knowing it they stood in much the same position as he had occupied some years before. The power of making electrical inventions, which Hopkinson had shared with very few others, came to belong to a much larger class owing its existence in large measure to him. The Edison-Hopkinson dynamo was of unique value because it was the first concrete expression of the correct principles of dynamo design. But it was a simple deduction from those principles, and after they were published in 1886 it was in the power of many to apply them in similar manner to the new needs which were constantly arising. So it was with the closed circuit transformer, and others of his inventions. It was for John Hopkinson to create the art of designing electrical machinery. The simpler matter of practising that art he soon left to others, and turned to work worthier of his powers.

Such work naturally followed the great extension of knowledge about electrical matters of which we have just been speaking. Hopkinson's advice to the Edison Company and other clients in early days had dealt with questions of general policy in electric lighting, without reference to particular circumstances—with model provisional orders, methods of charge and systems of supply. But with the spread of these and other ideas, there came into being numbers of actual commercial undertakings in electric lighting and power transmission, and as a necessary corollary the profession of consulting electrical engineer was established. Hopkinson took his place as leader of that profession. He came to be regarded as one whom anyone with an electrical project in hand must consult. Henceforward his professional work consists largely of advice on particular schemes.

In this way he influenced the course of most of the important electrical enterprises which were undertaken in this country during his lifetime, and was responsible for the execution of many. In London he was consulting engineer to the Metropolitan Electric Supply Company from its commencement, and he advised several other electrical corporations at critical stages in their careers. In the provinces he designed the Manchester Electric Lighting Works, and that undertaking was the first thing of the kind on a large scale which he carried out in detail. The works were designed in 1891 and were completed two years later. In Manchester Hopkinson had the satisfaction of putting into a concrete form several ideas which he had advocated for a long time. Chief among these was the maximum demand system of charging for electrical supply, which was here adopted for the first time. Another novel feature was the "five-wire"—an extension of the "three-wire"—method of distribution, the use of which was prompted by Board of Trade rules as to pressure of supply. Hopkinson foresaw the relaxation of these rules, and this led him to prefer the use of continuous current in many places where others with less foresight would have advocated the rival system—a prejudice (as some have thought it) which has been amply justified by experience. The success of the Manchester Station brought a good deal more work of the same kind in its train. Its details are of no great interest. Hopkinson himself grew tired of making electric lighting stations as their design became stereotyped, though he spared no pains in doing it. He took little pleasure in working on established lines; still less when they had been laid down by himself.

It is often said that England has been behind other countries in adopting improved means of transport. It must not be forgotten however that the City and South London Railway was constructed ten years ago by English engineers—at a time when nearly every other capital in the world had only horse trams. In the making of this railway John Hopkinson took an important part, acting as consulting engineer to Messrs Mather and Platt who were contractors for the generating machinery and rolling stock. In more recent developments of electric traction in this country he also had a large share. He was the engineer of all the more important city tramways on which electric haulage was adopted in his time. That from Kirkstall to Roundhay in Leeds, and the first electric lines in Liverpool, were constructed under his supervision. It is certain that had he lived he would have taken an even more prominent position in work of this class than in electric lighting construction. It must be admitted however that, in the particular case of electric surface tramways, the Americans had several years' start of us, and it may be doubted whether Hopkinson would have found much interest in problems which had already been pretty fully worked out by others. In constructive engineering he liked to do one big job of a kind, but did not care about the work of the same sort which would necessarily follow its successful accomplishment. A large practice in traction or lighting engineering had no attractions for him. He had one great desire, and that was to carry out a long-distance transmission of power. It would, he used to say, turn his hair white, but in doing it he would "fulfil the end of his being." It was a dream the realisation of which was probably only prevented by his death. He had been consulted on many occasions by people having such schemes in their minds. One of them—a big project for taking power from the Victoria Falls on the Zambesi—came near taking him to South Africa but was frustrated by the Matabele war. I well remember his joy at the prospect of doing this work; he would have thrown up everything for it. A year later a proposal for taking a large amount of power from Niagara to Toronto was brought to him. He did a great deal of work on it, and it was under consideration when he died.

The inspiration to purely scientific research which Hopkinson was wont to find in his professional work has been several times alluded to. With the change in the character of his business which has been mentioned above as having taken place about the year 1890 this stimulus disappeared to a great extent. The increasing calls on his time also left him less leisure for experiment, and these causes largely account for his lessened productiveness in this line during the last ten years of his life. The effects of both were mitigated to a great extent by his appointment in 1892 to the chair of Electrical Engineering which had just been founded by Lady Siemens at King's College, London. The experiments requiring the time which he could no longer devote to them personally could now be done under his direction by assistants, and the constant need to keep his students occupied with subjects of research was a new and powerful stimulus to producing ideas. A great deal of work which has already been noticed was done at the King's College Laboratory. The first experiments there were directed to the verification of the theory of commutation which John and Edward Hopkinson had enunciated in the 1886 paper on Dynamo Machines*. They were followed by the work on the alternating current dynamo in which a rotating contact maker was employed fixed to the armature, and making contact once in a revolution. By means of a quadrant electrometer the potential, whether of the machine itself or across a non-inductive resistance in series with it, could be measured at any given epoch. In this manner the potential of the machine, and the current passing in it, were plotted in terms of the time. The results disclosed the large effects of the currents induced in the solid cores by the varying action of the armature current†. Similar methods were applied to the determination of the change in the cyclic curve of magnetisation of iron which occurs if the cycle be effected 100 times per second or more, instead of quite slowly ‡. There was also an investigation of electrolysis with alternating currents, which led to a remarkable estimate of the size of atoms§. The last piece of work done at King's College was that on "Residual Charge and Capacity as affected by Temperature and Time." I think that this was in many ways my father's crowning achievement in pure research. Its place in his work has already been mentioned in speaking of the experiments done long before on Residual Charge, to which it was the fitting sequel. His con-

* Vol. I. p. 122.   † Vol. I. p. 156.
‡ Vol. II. p. 272.   § Vol. II. p. 383.

ception of "deviation from Maxwell's law" as only another name for residual charge found justification in these experiments twenty years after it was formed. I cannot give a better account of this paper than that contained in the abstract which was sent to the Royal Society Proceedings:

"The major portion of the experiments described in the paper have been made on window glass and ice. It is shown that for long times residual charge diminishes with rise of temperature in the case of glass; but for short times it increases both for glass and ice. The capacity of glass when measured for ordinary durations of time, such as 1/100th to 1/10th second, increases much with rise of temperature, but when measured for short periods, such as $1/10^6$ second, it does not sensibly increase. The difference is shown to be due to the residual charge which comes out between 1/50,000th second and 1/100th second. The capacity of ice when measured for periods of 1/100th to 1/10th second increase both with rise of temperature and with increase of time; its value is of the order of 80, but when measured for periods such as $1/10^6$ second its value is less than 3. The difference again is due to residual charge coming out during short times. In the case of glass conductivity has been observed at fairly high temperatures and after short times of electrification; it is found that the conductivity after 1/50,000th second electrification is much greater than after 1/10,000th, but for longer times is sensibly constant. Thus a continuity is shown between the conduction in dielectrics which exhibit residual charge and deviation from Maxwell's law, and (that in) ordinary electrolytes."

It is unnecessary to give here a catalogue of the honours and distinctions which were the natural results of Hopkinson's achievements. He cared little for such things and took no pains to get them. One or two however may be mentioned which did give him peculiar pleasure. Such were the Royal Society medal awarded to him in 1890 for his magnetic researches, and his election in 1886 and again in 1893 to the Council of the Society. That he should be President of the Institution of Electrical Engineers at an exceptionally early age was the natural consequence of his position, but his election to that post a second time in 1895 was a tribute to his personal character which he greatly appreciated, though the duties of President were little to his taste. It was at a critical period in the history of the Institution, and his fellow-engineers came to him not only as their acknowledged leader, but as one above all petty jealousy or desire for advertisement. Another honour which implied something more than mere scientific distinction was his election to the Athenaeum Club in 1887 as one of the nine annually chosen for eminence in "literature, art or science." Such things pleased him more than he admitted, though less than they would please most men.

This Memoir has dealt chiefly, as was fitting, with Hopkinson's professional and scientific work. It remains to say something of his activity in other directions. He had strong and original political views which he would advocate vigorously in conversation, though he had little ambition to impress them on a wider circle. He was however strangely moved by the events at the end of 1895 and beginning of 1896; indeed there is an entry in his diary at that time that he is unable to do much effective work in the present state of politics—the more significant as it is almost the only remark of a subjective kind which occurs anywhere either in his diary or in his later correspondence. His feelings in that crisis led him to the formation of the Electrical Engineers Volunteers Corps, of which he was the first Commanding Officer. He had the satisfaction of seeing the utility of the Corps proved, and its success assured, though it had hardly been long enough formed when he died to get the full benefit of his powers of organisation.

Of what is called "general reading"—such as novels, poetry, and philosophy—he did little in his later years. His favourite novelist was Thackeray; *Esmond* was a work he especially admired. He would also read Fielding on occasion. But George Eliot and Dickens did not attract him, still less any novelists of later date. Of poets he liked Tennyson better than any, doubtless because of his sympathy with modern scientific conceptions. I do not remember ever seeing him read a purely philosophical work though he had read many in early days and always liked to discuss philosophical or ethical questions. He regarded them as mental exercises, very pleasant, and no doubt healthful, but of no real importance. He had certain deep-seated moral convictions which argument could no more explain than it could destroy. A dishonourable action excited disgust; it was profitless, though amusing, to discuss why.

The most important part of Hopkinson's work outside his own profession lay, however, in education. From the early days when he instructed his brothers he was always a great teacher of those he loved. His method was to leave as much as he possibly could to the unaided intelligence of the pupil. It delighted him to watch a keen but untutored mind struggling with a difficulty, and he took care never to spoil the sport by too much assistance. From time to time he would give a hint or a suggestion just sufficient to put his pupil on the track again; only after long and painful floundering would he give a complete and consecutive demonstration. It was also "suppose you try so and so"; never "do you follow me?" I well remember my first introduction to the Calculus in this way. I was told to find the area of some curve, and for two hours or more I suffered all the troubles of the circle-squarers. Then, when I was pretty well beaten, a ray of light was admitted in the shape of a suggestion to try "cutting it up into little bits." More wrestling followed, then another hint, and so on till the problem was solved. On a small scale that was my father's theory of education in relation to the individual. He had a great belief in allowing young people, sound in body and mind, to go their own ways without let or hindrance. No obstacles should be put in the path, and guidance or help should only be given when it was really needed and when its value would be appreciated.

It was in his own family of course that John Hopkinson's influence as a teacher was deepest; but in his later years it was felt in a much wider circle. In addition to the great effects of his personality on the young men with whom he came in contact at King's College and in the course of his business, a large proportion of the electrical profession were latterly in the position of his disciples. It will be found that nearly all his later addresses to Engineering Societies were—if one may use the word without offence—of a didactic character: that is to say the subject-matter was not original, though put in a new way, and it was regarded by most of the audience as non-controversial. Twice in his life he came near to taking a University professorship; but he would not have been so powerful a teacher had teaching been his profession. No professor pure and simple, however eminent, could have spoken to electrical engineers with half the authority which he possessed. His utterances on points of pure science or mathematics might have been accepted—just as Hopkinson's were—without question; but his auditors would have felt doubts as to his sympathy with their aims. Someone would have been sure to say, directly or indirectly, that he was not a "practical man," with the implication that practical men must discount what he said. That cry was never raised against John Hopkinson, and therein lay the secret of his power as a teacher among engineers generally. In his classes at King's College, where he came personally into contact with individuals, the effect of his dual position in the world of theory and in that of practice was even more marked. He spent but little time there—on the average perhaps one hour a day—but he came there each time fresh from the execution of big engineering works. Sometimes he would turn at once to a question—such as the capacity of glass—purely scientific and so far as could be seen utterly without promise of practical utility. On other occasions he would bring an electricity meter or a storage battery to be tested. There were two cardinal points in his teacher's creed. One was that the educational value of a piece of work was in nowise impaired if it happened to be practically useful. "From an educational point of view it is just as useful to learn to thoroughly test a steam-engine as to determine accurately the value of V or of the ohm." The other, that it is worth a student's while even at a technical college to do pieces of purely scientific research. Both were constantly put into practice at King's. But the greatest lesson of all—more than any precepts which he could impart in the short half-hour at his disposal—was to see him at work, to see revealed the sources of his power.

Hopkinson took a very active interest in the more general aspects of education. As a member of the Senate of London University he was able to give expression to strong views on that necessary adjunct of teaching, examination. He had had as examiner and candidate an almost unique experience of it. One point on which he felt very strongly was that the number of examinations of much the same standard should not be multiplied. On this ground he had a great belief in the high function of the University of London, and I do not think he liked the idea of its forsaking its exclusively examining character, and becoming in part a teaching body. It was not a point on which he was disposed to insist very much but he certainly felt that London University was doing work which no other body ever would or could do, and doing it very well; work, moreover, which concerned the whole country and not London alone. That work could not continue if the University were at all controlled by the teaching staff of London Colleges; whether the advantages of such control would at all make up for what it destroyed was a point on which he felt considerable doubt. His opinion was of great weight, for he knew the education of engineers, who formed a large proportion of London University candidates and will form a larger proportion of its students, in every aspect most thoroughly; and though himself a London professor he was free from all suspicion of personal bias. Besides the experience of his own remarkable training he was closely concerned with engineering education as a member of the Council of the Institution of Civil Engineers. Then he advised the Syndicate whose labours resulted in the formation of the Engineering Laboratory at Cambridge. He took a very warm interest in that school from its commencement. It was the University of Cambridge in no small degree that made him what he was, and his opinions on her place in engineering education are the most fitting preface to this reprint of his papers. They were expressed in a Memorandum written for the Syndicate and were apropos of a proposal to add a properly equipped Laboratory to the old Workshops.

MEMORANDUM ON ENGINEERING EDUCATION.

[*Written for a Cambridge University Syndicate in* 1890.]

AN Engineering School in the University has two uses; first for men other than those intending to be Engineers, and second for those purposing to follow that profession. It is one of the greatest possible advantages to members of the Bar to have some knowledge of mechanics and mechanical operations. Many cases in the courts turn on mechanical points, and though it is wholly unnecessary for a barrister to be in any sense an Engineer it is almost a necessity if he is to do his work properly that he should have so far to come in contact with engineering appliances as readily to apprehend what they are, to understand the point of view of the practical men with whom he may have to deal in the case in hand, and to be able to understand the conventions which are used in mechanical drawing. Such a training as is required can be given in a variety of ways. A little time spent in a drawing office and in a workshop would almost answer the purpose by the mere presence of the student amongst mechanical appliances even if he acquire no skill as a handicraftsman. Work in an engineering laboratory would answer the same purpose though possibly less effectually. To the same class will belong men whose career in life will lie in managing one or other of the manufacturing industries of the country. Here it is unnecessary that a man should be capable of designing machinery, but it is most desirable that he should be competent to form an independent opinion upon the facts concerning machinery which may be laid before him. In the case of clergymen again it is desirable that they should be able to enter into the point of view of those with whom they have to deal, the more real the everyday duties of his parishioners can be to a clergyman the better will he be able to work amongst them. For this no profound knowledge is required but such knowledge as may be acquired by working for a few hours from time to time in a mechanical workshop.

For this large class then of those who have no intention of practising the engineering profession, workshops such as those at present established are in my judgment invaluable; properly managed they are calculated just to give that reality to knowledge which cannot be obtained from books and which opens up quite a different point of view from anything which the student is likely to obtain outside of the Engineering School or the School of Practical Physics. Whether or not the University should add a school for producing Engineers I think it should provide for imparting the education which I have indicated and which if not imparted at the University will not be given at all to those who should receive it.

For the second class of students, those who intend to become Engineers, different considerations apply; I do not think it probable that Cambridge will ever become a large school for Engineers. The majority of Engineers in the country must be practical men who can deal effectually with the actual execution of a limited class of work, and for these it is probably best that their workshop education should begin earlier than the usual time of leaving the University. But there is also room in the profession for others who have a wider education addressed to enabling them to deal with circumstances for which there is no direct precedent. For this I believe that the mathematical education of the University would be of great value; but with it should be coupled constant contact with the physical phenomena in some shape or other so that the men may realise that every bit of their mathematical knowledge has its counterpart in objective facts. Of course for such men the more nearly their physical work bears upon practical engineering the better. From an educational point of view it is just as useful to learn to thoroughly test a steam-engine as to determine accurately the value of V or of the ohm, but in learning to test a steam-engine the student learns a great deal that will be of value to him when he comes to practise his profession. In my opinion it is not necessary that the class of Engineers which the University is calculated to produce should be handicraftsmen, but they must at some time or other come in contact with the actual execution of work and will learn to know when work is properly done and when it is not. Probably this would be less effectually accomplished in a workshop such as the University at present possesses than in any reasonably good mechanical engineering workshop engaged in trade. If I were sending a son to the University who was afterwards to practise the profession of an Engineer I should not send him to the mechanical workshop as at present arranged but let him work at mathematics and in the physical laboratory. What I take it is really wanted is that in addition to the workshops there should be a physical laboratory addressed to dealing with those matters which have a particular interest to Engineers, such for example as the investigation of the mechanical properties of materials, the thorough testing of steam and gas engines, hydraulic experiments, the testing from every point of view of dynamo machines, but not their manufacture.

There is another reason why the University should be possessed of a properly equipped engineering school. Many men no doubt go to the University with talent for Engineering as yet undeveloped and who, if they were provided with an opportunity at the University, would probably go into that profession for which they were best suited, and who otherwise might be diverted into other professions. There are also probably many students who intend to pursue Engineering as a means of livelihood, but who desire to enjoy the general advantages afforded by the University in other ways than in their professional training. For both it is of great importance, and of great importance to the country as well, that they should have an opportunity to pursue at the University that line best adapted to their subsequent work.

If an Engineering Laboratory were provided for the University a workshop of some sort would be a necessity; for the laboratory alone the workshop might possibly be on a smaller scale than that which at present exists but in that case the workshop would be insufficient to deal with the class of students who did not contemplate becoming Engineers.

It seems to me then that in any case the University workshop should be retained as a most desirable part of the education of most of the men who have afterwards to deal with practical life. I think also that if funds are found, an Engineering Laboratory for the benefit of those who think of Engineering as a profession should be established under the same management as the workshops.

It has been suggested that a great deal of Engineering testing can be done with very simple appliances and at a small expense. This is perfectly true, but to do it at all you have to depend on the ingenuity and resource of your Professor, and in any case the appliances possessed by the University for teaching Engineering would compare most favourably with the appliances possessed by many colleges in the large towns.

It was suggested by Professor Reynolds that the University might raise subscriptions for the purpose of founding a proper Engineering Laboratory from outside of the University. I see no reason whatever why this should not be done, but if done I think the appeal should be made on the ground that in founding a professional school which might not attain to very large dimensions the University was also providing a very valuable element of education to many other students. There are but few Engineers who are members of the University of Cambridge, but there are hosts of men who must feel that a measure of mechanical training would have been of value to them in the career which they have selected.

1.

GROUP-FLASHING LIGHTS.

[Pamphlet first published in October, 1874.]

EXTENSION of trade, and the consequent increase in the number of Lighthouses upon frequented coasts, continually causes a demand for greater variety in the appearance of Lights, in order to avoid confusion from the nearness of Lights of the same character. This is apparent in the fact that the scheme proposed by the first French Lighthouse Commission, and intended to be complete, has subsequently required extension. The first French scheme admitted but three distinctions, the fixed Light, and revolving Lights with flashes every minute and every thirty seconds. Now, the French system comprises quick-flashing Lights, revolving Lights with red flashes alternating with white, and fixed Lights varied by flashes; of the last there are no less than twenty-three on the French coast. Indeed the use of red at all in revolving Lights, involving as it does a serious tax on the luminous power of the flashes, or increased expense for the same power, sufficiently indicates that new combinations are, and will continue to be required. Our present purpose is to offer in a complete shape two new forms of Lighthouse apparatus, and to point out the advantages they possess over some very useful forms now in use. Before doing so it will be well to examine what are the qualities of a good Light. That for any given cost the intensity of the light should be as great as possible, or conversely, that when a given intensity of light is required it should be attained at a minimum expense, is obvious. Most distinctions of beacons depend on the succession of intervals of light and darkness.

The following are suggested as rules of comparison of the efficiency of such distinctions :—

1. The Light must not be too long obscured or an accident might occur in the interval, which a sight of the Light would have prevented. What period of darkness is admissible is a nautical question, and will depend on the position the Lighthouse occupies and the nature of the traffic which it has to guide. Flashes at intervals of as much as three minutes have been in use, but the tendency is to prefer shorter periods, as in the case of South Stack, which is to be altered from flashes every two minutes to flashes every minute. We may therefore assert that, other things being the same, the efficiency is increased as the time of eclipse is shortened.

2. Unless the eclipse is very short it is necessary that the duration of the flash should be sufficient to take the bearing of the Light. It is this among other reasons which necessitates a special form of revolving dioptric apparatus for condensing the electric light; with the usual form for oil flames, the flash of a half-minute electric light would last but the fraction of a second. What time is sufficient for taking a bearing, is again a purely nautical question, but we may safely say that a flash of considerable duration is more useful than one which gives bare time for observation.

3. The character of the Light must not be too long in declaring itself, in other words the Light must pass through its phases in a reasonable period of time, indeed the shorter this period the better. The fixed and flashing Lights of the French system are usually characterised by a bright flash, preceded and followed by a very brief eclipse, occurring every three or four minutes; not less than that period of watching is needed to identify the Light. It is a question for those whose experience justifies an expression of opinion, whether in some circumstances such a length of time is not too much. In this respect the revolving Lights with red and white flashes combined, are less favourable than the ordinary revolving Lights of the same period; for example, to distinguish a half-minute revolving Light showing red and white flashes alternately, from one in which there is a red followed by two whites, requires a minute and a half. Suppose the navigator first sees a white flash, then a red and again a

white, a minute has passed and he must still wait thirty seconds to see if the next flash will be white or red. It should, however, be observed that this remark only applies to the hypothetical case in which two such Lights are placed near each other.

4. One point insisted upon by Authorities who have themselves had nautical experience is, that the distinctions should be as simple and easy to apprehend as possible. It is mainly on this account that the scheme proposed by Mr Babbage never received any practical recognition. For the same reason it is unwise to trust too much to any but very marked differences in the period of ordinary revolving Lights. A forty-five second should not be considered safely distinguishable from a minute flash.

5. The characteristic appearance of the Light must be maintained at all distances and in all states of the weather in which the Light can be seen at all. If red and white flashes are combined, the portion of light devoted to each flash must be such that they shall have equal penetrating power. It would appear that the fixed and flashing Lights so popular on the French coast, would lead to mistakes at times when the feeble fixed Light is obscured, if the intervals of the flashes were not so long as to distinguish them from the ordinary revolving Lights.

The additional source of variation which I now propose is, that revolving apparatus should be constructed to exhibit two or three white flashes in rapid succession, in place of one at stated intervals of time. This would increase the capacity for variation of the revolving Lights three-fold, we should have single, double and triple flashes at whatever intervals are now considered suitable for revolving Lights. The optical apparatus for producing such combinations would be simple and cost little more than an ordinary revolving Light. For the double flash of the first order we should require a twelve-sided Light, the axes of the panels being placed at unequal intervals, alternately 15° and 45° (Fig. A.); the effect of this would be, that, with an apparatus completing a revolution in three minutes, and using as source of light the usual four-wick flame, there would be a flash of about 2″ duration followed by about $5\frac{1}{2}''$ dark, again a flash of 2″, this double flash being separated from the next by about $20\frac{1}{2}$ seconds of darkness. The ordinary eight-sided revolving Light condenses 45° of the light of the flame into about 4° in azimuth, thus producing a flash of intensity 11¼, if the continuous light of the fixed apparatus be taken as unity. In the double flashing Light each panel has 30° in azimuth, and would therefore give to each flash an intensity represented by 7½. Let us see how far this form of Light fulfils the requirements of a good Light. With periods of half a minute the longest eclipse would be 20½ seconds, giving a slight advantage over a half-minute Light with eight sides, which is eclipsed for about 27 seconds. The two flashes near to each other would be almost as convenient for taking a bearing as a continuous flash lasting from the beginning of the first flash to the end of the second; we may therefore consider that we have 9½ seconds available to take a bearing. It would only be necessary to see the two flashes in succession to identify the character of the Light, 9½ seconds would suffice without any counting to

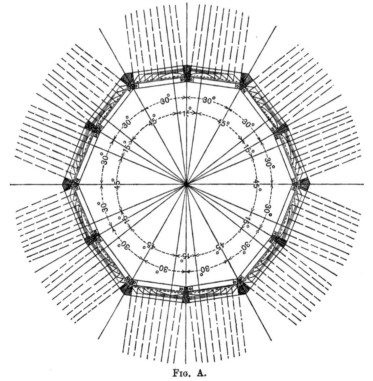

Fig. A.

recognise what the Light might be. The peculiarity of such a Light would lie, not in the precise periods between the flashes, but in the flash being a double one. No timing or counting is requisite, for the double and triple flash would be distinguished without any conscious process of counting. The two flashes of the pair being exactly of the same power, the general appearance of the Light must be always the same.

The triple flash would be conveniently obtained by a revolving apparatus of fifteen sides covering 24° each, the axes of the panels being placed at intervals of 48° and 12°. From such an apparatus, if a group of three flashes is to be exhibited every half-minute, each flash would last nearly two seconds, and the three would be separated by dark periods of three seconds. We should thus have a longest period of darkness of eighteen seconds, and twelve seconds in which to take a bearing. An apparatus giving four flashes in a group could be readily and economically formed of sixteen sides, there would not be the slightest difficulty in construction, and the flash would be of the same power as that of an ordinary sixteen-sided revolving Light. But it is doubtful if the need for counting so many as four flashes would not be found an unnecessary complication. All I would say here is that such a Light can easily be made.

The electric spark lends itself more readily than an oil flame to the production of any desired arrangement of the flashes of a revolving Light. Perhaps as unmistakable a form as any that could be suggested, would be a number of very quick flashes and eclipses, constituting a group which should recur at stated intervals.

It is worthy of notice that if a coast were lit on a system based on the use of group flashes, the appearance of the Light could be made to correspond to the blasts of the fog-horn or the strokes of the fog-bell; a group of three flashes, at intervals of thirty seconds, would naturally be used in conjunction with a fog-signal, sounded three times in succession at the same interval.

The following Table is intended to show the comparative advantages and disadvantages of group flashes, and the best forms of revolving Lights now in use. In this table the flashes or groups of flashes are supposed to have a half-minute period, the apparatus to be of the first order, and the divergence due to magnitude of flame to be 4°. The first column gives the power of the flash, the fixed Light being taken as unity.

	Power.	Time available for taking a bearing.	Time which may be required to identify the character of Light.	Greatest duration of darkness.	Percentage of whole Light wasted by use of colour.
Eight-sided revolving Light, of the usual form	$11\frac{1}{4}$	$2\frac{2}{3}$ Seconds.	30 Seconds.	$27\frac{1}{3}$	—
Sixteen-sided ditto	$5\frac{5}{8}$	$5\frac{1}{3}$,,	30 ,,	$24\frac{2}{3}$	—
Twelve-sided Light, giving a double flash	$7\frac{1}{2}$	$9\frac{1}{2}$,,	10 ,,	$20\frac{1}{2}$	—
Fifteen-sided Light, giving a triple flash	6	12 ,,	12 ,,	18	—
Nine-sided Light, giving a red flash, followed by two white	$6\frac{1}{3}$	3 ,,	90 ,,	27	36·84
Sixteen-sided Light, giving white and red alternate...	3	$5\frac{1}{3}$,,	90 ,,	$24\frac{1}{3}$	46·66
Eight-sided Light, red only...	$4\frac{1}{11}$	$2\frac{2}{3}$,,	30 ,,	$27\frac{1}{3}$	63·66
Sixteen-sided Light, red only	$2\frac{1}{22}$	$5\frac{1}{3}$,,	30 ,,	$24\frac{2}{3}$	63·66

It will be observed that in power, the double flash ranks second; in duration the group flashes are best, as also in respect to time required for identification.

Group-Flashing Lights can readily be obtained on the Catoptric system, or by using a number of small Holophotes, but such Lights would be subject to all the objections to which revolving Lights with single flashes produced by many lamps are obnoxious.

NOTE.—The Hopkinson Group-Flashing system was for the first time applied, in 1875, to the Catoptric Floating Light on the Royal Sovereign Shoals, near Beachy Head, and has since been applied to several Lightships of the Trinity Corporation.

The first Land Light on this system was for Tampico Lighthouse, Gulf of Mexico, Second Order triple-flashing, in 1875. Eighteen Sea Lights in all, and two Harbour Lights, have been constructed by Messrs Chance Brothers and Co., Limited, since 1875.

The Group-Flashing system has also been adopted by the French makers, who have, since 1876, supplied many lights to foreign Governments.

December, 1890.

2.

THE ELECTRIC LIGHTHOUSES OF MACQUARIE AND OF TINO.

[From the *Proceedings of The Institution of Civil Engineers*, Vol. LXXXVII. Session 1886–87. Part I.]

THE subject of the use of the electric light in lighthouses was fully discussed at the Institution in 1879, when Papers by Sir James Douglass, M. Inst. C.E., and by Mr James T. Chance, Assoc. Inst. C.E., were read*.

The subject has been further elaborately examined by Mr E. Allard†, and more recently in practical experiments, made at the South Foreland, exhaustively reported on by a Committee of the Trinity House‡. The justification of the present communication is that, at the lighthouses of Macquarie and of Tino, the optical apparatus is on a larger scale than has hitherto been used for the electric arc in lighthouses, and presents certain novel features in the details of construction. Further, as regards the electrical apparatus, tests were made upon the machinery for Macquarie when it was in the hands of Messrs Chance Brothers and Company, which still possess some value, although five years old; and, in the case of Tino, the machines are practically worked together in a manner not previously used otherwise than by way of experiment.

---

* *Minutes of Proceedings Inst. C.E.* vol. lvii. pp. 77 and 168.
† *Mémoire sur les Phares Électriques*, 1881.
‡ Report into the relative merits of Electricity, Gas, and Oil as Lighthouse Illuminants. Parts 1 and 2. PP. 1885.

In the case of both lighthouses, Messrs Chance Brothers and Company, of Birmingham, entered into a contract for the supply of all the apparatus required, including engines, machines, conductors, lamps, optical apparatus, and lanterns; and Sir James Douglass, Engineer-in-Chief of the Trinity House, acted as Inspecting Engineer to the respective Colonial and Foreign Governments.

As these two lighthouses present many features in common, it may be most convenient to give a full description of the earlier lighthouse, and then limit the description of Tino to those points in which it differs from Macquarie.

Macquarie.

This lighthouse is situated on South Head, near Sydney, the precise position being shown in a copy from the chart, Fig. 1. A lighthouse was first placed at this important landfall in 1817. The focal plane is 346 feet above the sea, and the distance of the sea-horizon is therefore 21·6 nautical miles, and the range about 27 nautical miles for an observer 15 feet above the sea.

Optical Apparatus.—The light is a revolving one, giving a single flash of eight seconds duration every minute. On account of the considerable altitude of the lighthouse, it was necessary to secure that a substantial quantity of light should be directed to the nearer sea; but it was also essential, on account of the exceptional power of the apparatus, that this dipping light should only be a small fraction of that sent to the horizon, otherwise its effect would be excessively dazzling. Many years ago, Mr James T. Chance urged that it was not wise to make use of very small apparatus for the electric arc, because a larger apparatus renders it possible for the optical engineer to effect with greater precision the distribution of light which is most desirable, and because any trifling error which may occur in the position of the electric arc has, with the larger apparatus, a less marked effect on the light as seen from the sea. In the lighthouses of Souter Point, the South Foreland, and the Lizard, the third-order apparatus of 500-millimetre focal length was adopted. Optically, the larger the apparatus used the better, but there might be some question whether, on purely optical grounds, the advantage of going beyond the third order is sufficient to justify the additional expense, but in the case of a revolving apparatus, the third order is a very inconvenient size for the service of the lamp; it is too large to be conveniently served from the outside, and too small to admit the attendant within it with comfort. With the large currents, which are now easily obtained and are likely to be used in lighthouses, a first or second order apparatus has the further

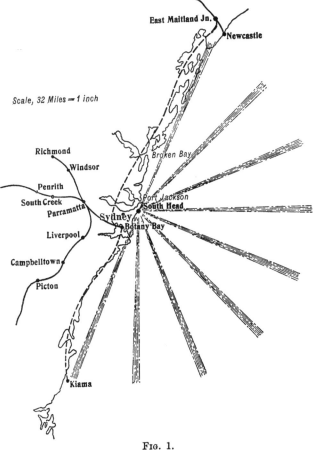

FIG. 1.

advantage that it is less liable to injury from particles thrown off from the heated carbons. In the case of Macquarie, it was decided to adopt an apparatus of the first order, 920-millimetre focal length; it was further decided that the optical apparatus should produce its condensing effect by means of a single agent; that is to say, the vertical straight prisms which were used in Souter Point and other revolving electric lighthouses should be dispensed with. The condensation and distribution of light necessary may be obtained by means of a single agent, with apparatus such as has been proposed by Mr Alan Brebner, jun., Assoc. M. Inst. C.E.*; but this construction is open to the objections that it is somewhat costly, and that it increases the length of the path of the rays through the glass, and consequent absorption. A practically better plan is to adopt forms not differing very greatly from those introduced by Fresnel; to specially arrange them for the purpose in hand, and to accept certain consequent minute deviations from a mathematically accurate solution for the sake of advantages of greater importance, when all the actual conditions are taken into account. Fig. 2 shows the optical apparatus in vertical section; the upper and the lower totally-reflecting prisms are, as is usual in revolving lights, forms of revolutions about a horizontal axis; they direct the light incident upon them to the horizon and the distant sea from 10′ above the horizon to 30′ below; they are specially adjusted to distribute the light in azimuth over the arc of 3° necessary for a proper duration of flash.

The refracting portion of the apparatus has the profile so calculated that the central lens, and the three rings next to the lens above and below, direct their light to the horizon without vertical divergence, except what is due to the size of the arc; the light for the nearer sea is obtained from the remaining ten lens-segments, Nos. 5 to 9 inclusive, above and below the centre, counting the centre as No. 1, the distribution being according to the following Table, in which the first column gives the denomination of the elements of the lens in accordance with the numbers marked upon the section; the second, the angle between the direction of the sea-horizon and the ray emerging from the upper limit of the element; the third, the angle between the direction of the sea-horizon and the ray from the lower limit of the element, the negative sign denoting that the emerging ray

* *Minutes of Proceedings Inst. C.E.* vol. lxx. p. 386.

is above the horizon. This practice, of appropriating certain elements of the apparatus to different distances on the sea, was

I.	II.	III.
9 top	− 10 0	2 30 59
8 ,,	2 30 59	5 8 52
7 ,,	− 10 0	2 37 30
6 ,,	− 10 0	1 30 0
5 ,,	− 10 0	1 0 0
5 bottom	− 10 0	1 0 0
6 ,,	− 10 0	1 30 0
7 ,,	1 30 0	3 44 27
8 ,,	3 44 27	5 50 41
9 ,,	5 50 41	7 46 57

first introduced by Mr James T. Chance, in the lights of the South Foreland exhibited in January 1872.

The ray, dipping at an angle of 7° 46′ 57″ below the horizon, will strike the sea at ½ mile, while 5° 8′ 52″ corresponds to ¾ mile, 2° 37′ 30″ to 1¼ mile, 1° 30′ to 2 miles, 1° to 2½ miles, and 30′ to about 4 miles. Thus the direct light begins at about ½ mile from the lighthouse. From ½ mile to ¾ mile the sea receives light from one element of the apparatus, from ¾ to 1¼ mile from two elements, from 1¼ mile to 2 miles from three elements, from 2 to 2½ miles from four elements, and beyond 2½ miles from six elements; the upper and lower totally reflecting prisms come in aid at about 5 miles. The main power of the apparatus is hardly attained till a distance of 8 or 10 miles. Fig. 3 is a sectional plan of the apparatus by a horizontal plane through the focus. It will be seen that a dioptric mirror is placed on the landward side of the arc. This mirror is arranged to form the image of the arc at one side of the carbons, so avoiding the interception of light which would result if the mirror were used in the ordinary way, and contributing to the horizontal divergence necessary. Further horizontal divergence is given by the form of the lens. In the ordinary revolving light the inner face of the lens is plane; here it is cylindrical, the axis of the cylinder being vertical. This method of obtaining horizontal divergence is a modification of a proposal of Mr Thomas Stevenson*, M. Inst. C.E.; it is not mathematically accurate, inasmuch as the cylindrical form of the inner

---

* *Lighthouse Construction and Illumination*, p. 186.

face of the lens not only displaces the emergent ray horizontally, but also in the case of rays not in the vertical nor horizontal plane

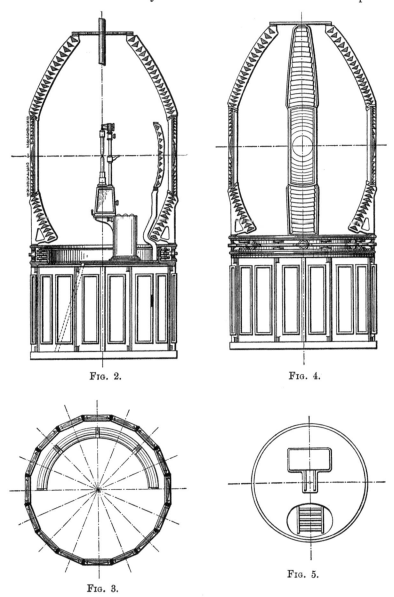

Fig. 2.   Fig. 4.

Fig. 3.   Fig. 5.

through the focus, to a small extent vertically, but the error is easily calculable, and is unimportant, provided the lens is narrow, and the horizontal divergence of the beam moderate. Plate 9, Fig. 4, shows a complete panel in elevation with revolving carriage. Fig. 5 shows the plan of the service-table of the pedestal and lamp-table. A new construction was adopted for the gun-metal framework of the optical apparatus to reduce the interception of light by the frame to a minimum. The metal segment A, Fig. 6, forms part of the lower prism-frame, B part of the upper frame, whilst C and D are parts of the frame for the refracting portion of the apparatus; uprights E support the upper prism-frames without throwing weight on the lens-frames. With the ordinary constructions of frame, Figs. 7 and 8, the equivalent of these ring segments A and B would intercept about double as much light as in this new construction.

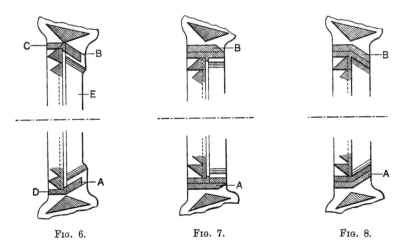

FIG. 6.　　　　FIG. 7.　　　　FIG. 8.

Mechanism for Rotation.—The pedestal is similar to those designed by Sir James Douglass to permit the light-keeper to obtain access from below to the interior of the apparatus without in any way interfering with its rotation. The clock-work is fitted with the governor, and maintaining power used by Messrs Chance Brothers and Company for the last twelve years. The roller-ring may be mentioned as of an improved type, for although it has been used for some years in all Messrs Chance's lights, it has not been described before. The rollers and roller-paths which carry the whole weight of the optical apparatus have long been made conical, so that the surfaces roll upon each other without twisting. There is consequently a very considerable radial force on each roller tending to force it outwards, the reaction against this force causes a very important part of the total frictional resistance. Fig. 9 shows a portion of the roller-ring and one of the conical rollers, according to the old construction; Figs. 10, 11, according to the improved construction; in the former it will be observed that the thrust of the roller is received on a collar; in

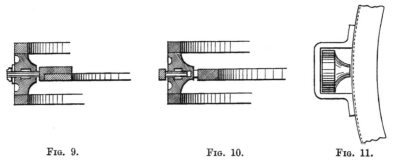

Fig. 9. Fig. 10. Fig. 11.

the latter, on the end of a pin. The reduction of friction is practically very considerable, and although of small importance in a slow-moving apparatus like Macquarie, is of great importance in heavier and quicker apparatus; for example, the triple-flashing light at Bull Point, in Devonshire.

Lamps.—These are of the Serrin type, and were supplied by Baron De Meritens.

Lamp-Table.—The arrangements for rapidly changing electric lamps, and for substituting gas or oil when desired, are shown in Figs. 12, 13, 14, 15.

The intention was to use a gas-lamp in clear weather, and half-power or full-power electric light in thick weather, according to the opacity of the atmosphere; but the Author understands that in practice the electric arc is always used. The paraffin oil-lamp is intended as a resource in case of failure of the supply of gas.

Focussing the Arc.—Two approximately rectangular prisms are fixed upon the mirror frame at about 90° from each other, the longer face of each is plane, the other two faces convex, of such curvature as to form a good image of the arc upon the service-table, as shown in Fig. 16. During daylight, a pointed sight or focimeter is placed at the position of the image formed by the lens of an object on the horizon; this then is the position which the arc should occupy. A sight is next taken over the focimeter into one of the adjusting prisms, and a bright object such as a threepenny piece placed on the service-table, is moved about until its centre is seen in the prism, exactly upon the point of the focimeter; a mark is made in the then position of the object.

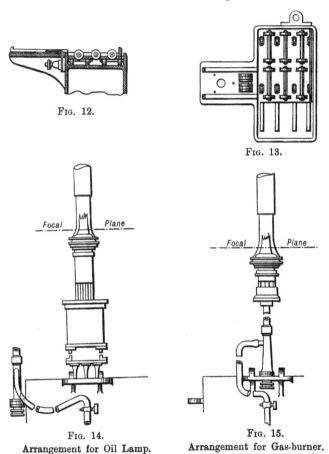

FIG. 12.

FIG. 13.

FIG. 14.
Arrangement for Oil Lamp.

FIG. 15.
Arrangement for Gas-burner.

When the arc is correctly adjusted, its image on the service-table will be at the point where the mark is made. Two prisms are used in order to secure that the arc shall be in the centre of the apparatus as well as at the correct level.

Lantern.—The lantern is of the well known Douglass type*.

* *Minutes of Proceedings Inst. C.E.* vol. lvi. p. 77.

Dynamo-electric Machines.—Two alternate-current machines, with permanent magnets manufactured by De Meritens, were supplied. Each machine has five rings in its armature, and in each ring there are sixteen segments. In supplying one arc for a lighthouse the machine runs about 830 revolutions per minute, and gives a current of 55 amperes when half the coils are used,

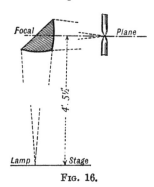

Fig. 16.

and of 110 when the whole of the machine is in action, the internal resistance in the two cases being 0·062 and 0·031 ohm. It is unnecessary to give a description of the machine as its general construction and dimensions are well known, but some numerical details are given below.

Engines.—Each machine is driven by an 8 H.P. Crossley gas-engine through a belt without countershafting.

Tests.—Whilst the dynamo machines were at the works of Messrs Chance Brothers and Company, a series of experiments was made in March 1881 to determine their properties. The time is long passed when it would be profitable to give the details of these experiments, but the general conclusions drawn at the time are still interesting. When the external resistance was a metallic conductor with small self-induction, it was found that with varying resistance and speed the currents observed agreed fairly well with calculation from the formula

$$\frac{A}{\sqrt{R^2 + \left(\frac{2\pi\gamma}{T}\right)^2}},*$$

---

* Lectures on the "Practical Application of Electricity." Session 1882-83. *Some points in Electric Lighting.* By Dr John Hopkinson, p. 88.

in which R is the total resistance of the circuit, γ the self-induction, and T the periodic time. When the machine was running 830 revolutions per minute $A = 67$ volts and $\left(\frac{2\pi\gamma}{T}\right)^2 = 0\cdot 197$ in ohm squared, hence $\gamma = 6\cdot 4 \times 10^5$ centimetres. The eighty sections of the machine are arranged four in series, twenty parallel. For a single section the value of γ would be 32×10^5 centimetres. The maximum induction in the core, which has an area of 5 square centimetres, is 24,600 or 4,920 per square centimetre. The loss of power was greater when the machine was doing little or no external work than when that work was great. This is clearly seen in the following table:—

Current ampères	7·70	73·60
Electrical work H.P.	0·69	5·66
Mechanical work applied	3·09	6·55
Loss	2·40	0·89

Photometric experiments were made upon the arc, and simultaneous measurements of effective power applied and of current passing. The red light was measured through bright copper ruby glass, and the blue through a solution of sulphate of copper and ammonia. The H.P. was measured by a transmission dynamometer; but the results must be accepted with some reserve, on account of the difficulty of ascertaining the mean tension in a strap which is constantly varying. The oscillations of the dynamometer were damped by a dashpot containing tar.

	Half Power.	Full Power.
Red candles	1,988	4,708
Blue „	4,079	11,382
Current (ampères)	54·5	105
Mechanical power applied (H.P.)	4·5	6·9
Power expended in heating conducting wires (H.P.)	0·24	0·95

The results illustrate the fact that, as the current increases, the total light increases in a higher ratio, red light in a slightly higher ratio, and blue in a considerably higher.

The machinery for this lighthouse was sent out to New South Wales in November 1881, and was put up and started under the superintendence of Mr J. Barnett, the Architect of the Colony, to whom is mainly due the success of the whole from the first start. The glare of the light upon the sky is said to have been seen at a distance of over 60 miles, far beyond the distance at which it would cease to be directly visible. The only criticism from mariners has been that when somewhat near the lighthouse the flashes are so bright as to dazzle the eye. This is an excellent proof of the power of the light, as a much smaller proportion of the light is directed upon the nearer sea than in any previous lighthouse. The lesson is that with powerful electric lighthouses almost all the light should, in ordinary weather at least, be directed to the horizon, and that the quantity thrown upon the nearer sea must be strictly limited. This is only possible when the focal length of the apparatus is large.

TINO.

This station is on a small island at the mouth of the Gulf of Spezia. Fig. 17 is copied from the chart of the neighbourhood. The focus is 386 feet above sea-level. The distance of the sea-horizon is 22·7 nautical miles, and the range practically 28 miles. The conditions, therefore, were very similar to those of Macquarie, with the exception that it was required to throw some light down into the channel between Palmaria and Tino. The

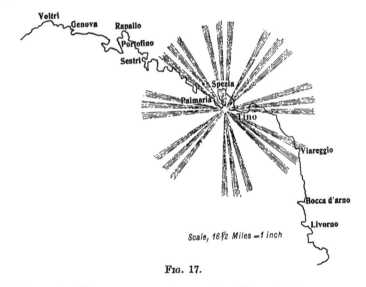

FIG. 17.

lighthouse itself presents some interesting historical features. The buildings were originally a place of defence against the pirates who occasionally made descents upon the coast. Subsequently a coal-fire lighthouse was established, and in the spring of 1885 part of the stock of lignite was still found to be in some of the buildings, where it had been lying for fifty years. In 1839 a dioptric light was established, one of the earliest of Fresnel's types, the lens-ring being replaced by short straight prisms, which formed by no means a bad approximation, and could be ground without special machinery. The present electric lighthouse has been in contemplation for several years.

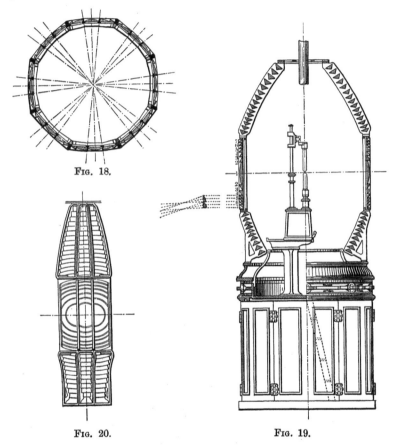

Fig. 18.

Fig. 20.   Fig. 19.

Optical apparatus.—The distinctive character of the light is a triple flash every half-minute. The apparatus for producing this effect is of the general form introduced by the author in 1874. In October of that year he issued a pamphlet pointing out the several advantages of group-flashing lights, showing for the first time a simple dioptric apparatus suitable to their production, and also pointing out how easy it is to give the group-flashing effect with catoptric apparatus. Since that time a large number of dioptric group-flashing lights have been made by Messrs Chance Brothers and Company, and some also in France, and Mr Allard has incorporated group-flashing lights in the system of distinctions he recommends; also a considerable proportion of the light-vessels on the English coasts have been converted into group-flashing lights of the catoptric system. On the ground of economy the second-order apparatus of 700-millimetre focus was adopted in the case of Tino. It is just large enough for tolerably convenient service of the lamp by an attendant entering within the apparatus. The apparatus, shown in vertical section in Fig. 18, and in horizontal section through the focus in Fig. 19, has twenty-four sides, eight groups of three; one group of three is shown in elevation in Fig. 20. The horizontal divergence is obtained in exactly the same way as at Macquarie, excepting that no mirror is used. The metal framework, however, approximates to the ordinary type, as the type used at Macquarie would have been costly when applied to a triple-flash light. The distribution of light vertically is as follows: upper and lower prisms, and the central lens, with the two lens-rings next adjoining it, all to the horizon and most distant sea. The lens and lens-rings direct their rays according to the following Table, which is arranged in exactly the same way as the Table already given for Macquarie—

I.	II.	III.
	° ′ ″	° ′ ″
7 top	0 31 35	3 16 0
6 ,,	..	2 0 0
5 ,,	..	1 30 0
4 ,,	..	1 0 0
4 bottom	..	0 45 0
5 ,,	..	0 30 0
6 ,,	..	0 30 0
7 ,,	all to the horizon.	

No. 7 bottom was directed wholly to the horizon, in order to avoid the horizontal bar of the lantern. It will be observed that the quantity of light thrown upon the nearer sea is much less in the case of Tino than in that of Macquarie, and that greater reliance is placed upon the accuracy with which the arc can be kept in focus; experience has justified these changes, as improvements of a perfectly safe nature.

A small part of each flash is bent downwards and distributed over the channel between Tino and Palmaria, by means of subsidiary prisms fixed upon the lantern, shown at X, Fig. 18. These subsidiary prisms are really superfluous, as the scattered light from the beams overhead is found to be as effective at this short distance. Fig. 21 shows the plan of lamp shunting-table and service-table.

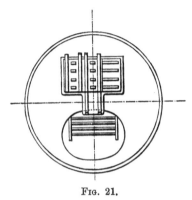

FIG. 21.

Engines.—As there is no water upon the island, the practice of the Trinity House was followed, and two of the Brown hot-air engines were supplied, each driving through a countershaft one of the machines. The countershafts could be connected by means of a Mather and Platt friction-coupling, so that the two machines could be driven together, or either machine from either engine. Drawings of the Brown engines are given in Sir James Douglass's Paper*. The accompanying indicator-diagrams were taken from the compressing- and working-cylinders. Whilst these diagrams were taken the effective power developed was measured by a friction-brake on the driving-pulley, and was found to be 9·1 H.P. Thus of 33·1 H.P. indicated in the working-cylinder, 17·7 H.P. is employed in compressing the air, 6·3 H.P. is wasted in friction in various parts of the machine, and only 9·1 H.P. is effective upon the brake. The engines consume about 4 lbs. of coke per effective H.P. per hour. In future lighthouses, when a steam-engine cannot be employed, it would be preferable on every ground to use

* Minutes of *Proceedings of the Inst.*, vol. LVII. Plate 6.

gas-engines, and manufacture on the spot either Dowson gas or ordinary gas, according to the character of the fuel available.

Dynamo Machines.—There are two machines of exactly the same type as those supplied for Macquarie, the only novelty lying in the method of using them. In 1868 Mr Wilde discovered, by experiment, that two alternate-current dynamos, independently driven at the same speed, would, if electrically connected, so control each other's motions that they would add their currents. The

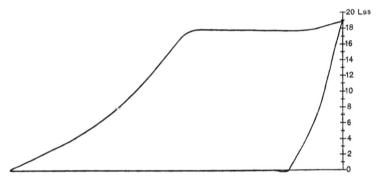

Fig. 22.—Compressor-pump cylinder, 24 inches in diameter. Stroke, 22 inches. Indicated H.P., 17.7.

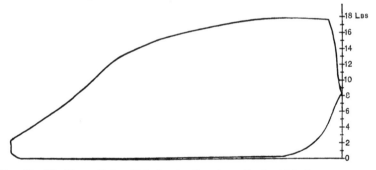

Fig. 23.—Working cylinder, 32 inches in diameter. Stroke, 20 inches. Indicated H.P., 33.1. Revolutions per minute, 64. Power on brake of fly-wheel, 9.12 H.P. Pressure in reservoir, 19 to 24 lbs.

author subsequently arrived at the same conclusion independently, on theoretical grounds, and gave a thorough explanation of the fact[*]. The result has been put to a practical application at Tino.

[*] Lectures on the "Practical Application of Electricity." Session 1882–83. Paper on "Some Points in Electric Lighting," reprinted in this volume. By Dr John Hopkinson. And *Journal of the Society of Telegraph Engineers and Electricians*, vol. XIII. p. 496.

The machines are connected to a single switchboard, so that each half of the two machines can at pleasure be connected to, or disconnected from, the main conductors. Thus a current can be supplied from either machine at half power, 55 ampères, or full power, 110 ampères, or from the two machines of double power, or about 200 ampères. Further, a change can be made without extinction of the light from one dynamo and engine to the other. Thus, suppose one machine is working full power, clutch the countershafts gradually together, so starting the second engine; throw on the band of the second machine, cut out half the first machine, and connect half the second machine at the switchboard; the two machines at once synchronize, without affecting the light. Disconnect the remaining half of the first machine, and connect the remaining half of the second, unclutch the countershafts, and stop the first engine. One man can effect the change, with no more disturbance of the light than a change from full to half power for about one second. A further conclusion, deduced from theoretical considerations, was that of two alternate current machines of equal potential, one could be used as a generator of electricity, the other as a motor converting the current generated back into mechanical power. It was found impossible to verify this conclusion with such intermittent driving as that of a hot-air engine. But Professor W. G. Adams effected the verification without difficulty at the South Foreland, the motive power being steam.

Lamps.—These are the improved Serrin of M. Berjot. One of the three lamps supplied is of larger size, for the double power current from the two machines. This lamp was said to be suitable for a still greater current, but with about 200 ampères it soon became dangerously heated; a simple modification rendered the lamp equal to the actual work it had to do. It is, however, probable that for the occasional circumstances when it is necessary to use so great a current as 200 ampères in a lighthouse, a lamp worked partly by hand would be preferable to a regulator entirely automatic.

The apparatus was delivered in November, 1884, and was put up by workmen from Messrs Chance's workshops, under the supervision of Mr L. Luiggi, of Genoa, to whose ability and energy the complete success of the lighthouse is largely due. A complete test of the performance of the light, as seen from the sea in all grades of its power, was made in April, 1885, by a commission, consisting of Professor Garibaldi, of Genoa; Mr Giaccone, engineer-in-chief for Italian lighthouses; Captain Sartoris, and Mr Luiggi, the author attending on behalf of Messrs Chance. The light was well observed through rain, when distant 32 nautical miles, and although below the horizon, the position was precisely localized, and the triple flash distinction unmistakeable. At 18 miles distant the illumination of the flash upon white paper was sufficient to make out letters marked in pencil $1\frac{1}{4}$ inch high, and when 14 miles distant it was easy to ascertain the time from a watch. The light is frequently seen at a distance of 50 miles near to Genoa.

A review of work which has been carried out naturally suggests many questions as to what conclusions experience has established, and what indications it gives of the probable direction for future developments. In the use of electric light in lighthouses there are many questions upon which there is wide difference of opinion, questions both as to when and where electric light should be adopted, and questions as to the best way of employing it. It may not be unprofitable to allude to some of them. Although English engineers are now well agreed that a large optical apparatus should be used for the electric light, this opinion is not universally accepted. The advantages of a large apparatus have already been mentioned. To balance them, there is nothing on the other side but the less prime cost of the smaller apparatus. Although the difference of cost appears considerable when attention is confined to the optical apparatus, it is unimportant when the whole outlay on the lighthouse is brought into account. Cases are, however, conceivable in which a small optical apparatus such as a fourth order, having a focal distance of 250 millimetres, would be properly preferred; such, for example, as a harbour light which could be supplied with current from machinery also used for other purposes, but such cases are likely to be exceptional.

When a flame from oil or gas is the source of light, there is of necessity a considerable divergence vertically; and the distribution of the light through the angle of vertical divergence is not at disposal, except to a very limited extent in some cases, but is determined by the size and character of the flame. With the electric arc and a large optical apparatus it can be determined in considerable measure how the light shall be distributed—how much shall be sent to the distant sea, how much to the various distances between the foot of the tower and a distance of some miles. It becomes then a question what use is to be made of this facility. The experience at Macquarie and at Tino is emphatic, that it is in every way advantageous to direct much the greater part of the light to the horizon with a very small divergence, and to distribute the comparatively small remainder over the nearer sea with intensity increasing with the distance.

A question allied to the last is this: Whether it be desirable to provide means of directing the strongest light downwards on to the nearer sea in time of fog? The answer must depend upon the circumstances of the particular locality. Take the case of a lighthouse on an isolated rock, the purpose of which is primarily to be a beacon to keep ships off that rock; a lighthouse which would not exist were it not more practical or cheaper to build and maintain the lighthouse rather than remove the rock. Here surely it is of the greatest use to provide means whereby, if the light cannot penetrate 2 miles, it shall if possible be visible at 1 mile. But other cases occur in which the lighthouse has to cover a long length of coast, and has almost as much to do with points of the coast 10 miles distant as with the point upon which it is placed, cases in which the lighthouse is far more useful in guiding the regular traffic passing within a radius of 20 miles or more than in preventing vessels running ashore within a mile of the tower. Such a light fails of its purpose if it can only be seen at a distance of a mile, covering less than $\frac{1}{400}$ part of its normal area of illumination; it becomes comparatively useless unless it penetrates to something like its normal range, and its efficiency must be measured by the fewness of the occasions when it fails to do this. It is a grave question whether it be prudent in such cases to place upon the light-keeper the responsibility of judging when the light should be dipped on to the nearer sea, the fact being that, if his judgment errs, he may actually diminish the range of the light, and cause unnecessarily the lighthouse to fail of fulfilling its most important function. It is easy for him to be misled if the fog is local and does not extend to any great distance from the lighthouse. Another element enters into the consideration—the height above the sea. If the focus be 100 feet above the sea-level, the dip of the sea-horizon is 9′ 45″, and a ray dipping 9′ 45″ below the sea-horizon will meet the sea at a distance of 3·1 nautical miles from the tower. Even with a first-order apparatus, if the arc be a powerful one, it is very difficult to render the light directed to the horizon from an elevation of 100 feet more powerful than that directed to a point distant 4 miles from the tower. Unavoidable divergence will render the two intensities practically equal.

Passing to questions of another class, what are the relative advantages in an electric lighthouse of continuous and alternating currents? Present practice tends altogether in favour of alternate currents, but this practice largely results from unfavourable experience of the older continuous-current machines. These machines have in many respects been greatly improved in the last two or three years. The continuous current presents the advantage of greater economy of power in producing the current, less floor-space required by the machine, and a smaller prime cost. The alternate-current magneto machine, on the other hand, has the advantage that it may be driven with a defectively governed prime mover, with an indifferent lamp, and may suffer neglect with impunity; whereas the more compact and efficient continuous-current machine would be in serious peril of destruction. Optical apparatus can be constructed suitable to make the most of either form of arc. Hot-air engines have found favour for electric lighthouses, because in many cases there is no available supply of fresh water. The engines of which the author has experience are open to the objection that they take a great deal of room, are not economical of fuel, and do not govern so quickly as is desirable; the wear and tear also, when they are worked to anything like their full power, is very serious. A gas-engine, with Dowson or other gas made on the spot, could be used with greater advantage.

Antecedent to all considerations as to the best apparatus and machinery to be used, is the question, under what circumstances, if at all, should electric light be used in a lighthouse? The Trinity House experiments at the South Foreland showed to demonstration that, where the issue to be decided was how to produce a light which should be capable of penetrating the furthest in all weathers, electric light could do that which could be done in no other way, and that it was the cheapest light of all when the price is estimated per unit of light. But the conclusion was also reached that an electric light must inevitably cost a large sum, both in first outlay and in maintenance; therefore that electric light is extravagant unless very extraordinary power is a necessity. This conclusion is doubtless a fair consequence of experience, but it is not an inherent property of electric light. Both the capital outlay and the cost of maintenance are greatly increased by the practice of so arranging the machinery as to provide, at all times, a light of very great power: whence follows that the machinery must be placed at some distance from the lantern, and two men must always be on duty; one man in the lantern, and another with the machinery.

The essentials for a cheap electric lighthouse are, that for ordinary states of the atmosphere there shall be provided a plant under the easy control of the light-keeper himself, which shall be precisely adapted to produce that amount of light which is wanted in ordinary states of the atmosphere; but for thick weather there shall be provided a much more powerful engine and dynamo, available also as a reserve in case the smaller machinery from any cause breaks down. The occasional machinery may be more remote from the lantern, as it is a small matter to require a second man to work on the comparatively rare occasions when the maximum power is needed. A small gas-engine and a dynamo machine can be placed without any crowding in the room immediately below the lantern, and arrangements can be made whereby the light-keeper, whether he is in the lantern or in the engine-room, can ascertain at a glance whether the arc is in its proper position, with an error of less than 1 millimetre. The attendance on the lamp, rotating apparatus of the lens (if a revolving light), engine and dynamo, would be easy when the whole is brought together, so as to be under observation at once; in fact the gas-engine, dynamo, and lamp constitute together a gas-burner which, though consisting of many parts, is automatic throughout, and requires nothing but the constant presence of a custodian, exactly as the gas-lamp in a lighthouse requires a custodian, as a guarantee against failure. The same end, viz., the concentration of the whole mechanical and electrical apparatus under one pair of eyes, could be attained, of course, in other ways. Accumulators could be used, or a petroleum-engine.

In order to give definiteness and afford facilities for criticism, the better course will be to describe a suitable machinery; state what it will do, what attendance it will require, and what it will cost. The author proposes, then, for an electric lighthouse where small outlay is essential, the following:—A Dowson gas-producing apparatus and gas-holder, the generator and superheater being in duplicate, each capable of making 1,200 feet of gas per hour, the gas-holder having a capacity of 3,000 cubic feet.

An 8-H.P. nominal Otto gas-engine and series-wound dynamo-machine, placed in a room near the base of the tower, and copper conductors to the lantern, the dynamo having magnet coils, divided into sections so as to supply a small current when required.

A 1-H.P. nominal Otto gas-engine and dynamo-machine, placed in the room immediately beneath the lantern floor, with gas-pipe from the gas-holder; three electric lamps, to receive either carbons 25 millimetres in diameter or any lesser size, with complete adjustments for accurate focussing; one paraffin lamp as a substitute; an optical apparatus of the second order of 70-centimetre focal distance. The cost of this apparatus would depend upon the character of the light it was intended to exhibit. To fix ideas, let it be assumed that the light is to be a half-minute revolving light, showing all round the lighthouse. There could then be supplied a sixteen-sided apparatus with pedestal and revolving machinery. Provision would be made in the optical apparatus for giving the horizontal and vertical divergence desired by the same methods successfully used in the lighthouses of Macquarie and of Tino.

Two focussing prisms would be fixed to form magnified images of the arc, on pieces of obscured glass let into the pedestal floor, so that the keeper, whether in the lantern or in the engine-room, could see at a glance the state of the arc, and observe whether it is of proper length with the carbons in line, whether it is exactly at the right height and in the centre of the apparatus. An error of 1 millimetre would be glaringly apparent, and call for immediate adjustment, although its effect would be only a displacement of the beam 5' of angle.

The lantern would be 10 feet diameter, with bent plate-glass.

The cost of the whole above described would be materially less than the cost of a first-order light and lantern with oil-lamp and large burners.

Now what result would be obtained? In fine weather the small engine would be used. Its effective power on the brake is fully 1¼ H.P.; from this 1½ H.P. the dynamo-machine produces considerably over 800 watts, say 800 watts in the arc itself, or 20 ampères through a fairly long arc of 40 volts. Of course the value of this in candles depends upon the colour in which it is measured, and the direction in relation to the axis of the carbons. In red light the mean over the sphere would certainly exceed 1,200 candles. In clear weather or in slight haze or rain, the beam of this light through the lenses would be much more powerful at the horizon and on the more distant sea than any single-focus light with oil or gas as the illuminant, and would at least be fairly comparable with anything yet exhibited with oil or gas, whether triform or quadriform. But on the nearer sea the illumination would be reduced, so that no annoyance would be caused by dazzling flashes. In thick weather or indeed in any weather when there was a doubt as to the visibility at the horizon of the lower power, the large engine would be used under the superintendence of the second keeper. This engine will give 10 H.P. on the brake, and there is no difficulty in obtaining 85 per cent. of this as useful electrical energy outside the machine, that is, 6,340 watts. From this deduct 10 per cent. for the leads and the lamp and for steadying the arc, leaving 5,710 watts in the arc itself, or 114 ampères, with a difference of potential of 50 volts. Having regard to the fact that the optical apparatus here proposed acts upon a larger portion of the sphere than that used in the South Foreland experiments, that the vertical divergence is less, and that the potential difference is greater and the current continuous, although less in quantity, it may safely be assumed that the power of the resulting beam would not be inferior. It hence follows, from the South Foreland experiments, that in any fog the flashes would penetrate farther than those of any existing gas or oil light. The increased size of crater, compared with that produced by the current of 20 ampères, will give increased vertical divergence, and so cause the maximum illumination to be attained at a less distance from the lighthouse. The attendance of two men would suffice for all the duties of the lighthouse, because under ordinary circumstances one man only need be on duty excepting for two to three hours whilst gas is being made. The consumption of coal would be 4 lbs. per hour of lighting, of water about ½ gallon, of carbons about 4 inches. The whole cost of maintaining the light would differ little from that of an ordinary oil light of the first order.

Though it be the fact, that it is possible to exhibit an electric light at moderate cost, it does not follow that it is suitable for all ocean-lights. There is no room in a rock lighthouse tower for a gas-plant, and few would at present be prepared to recommend a petroleum-engine burning oil of a low flashing point. The light-keeper again must understand a gas-producer, a gas-engine, a dynamo, and an arc lamp, instead of only a paraffin lamp and burner, and arrangements must exist for repairing the more extensive machinery. Such considerations will justly weigh against the use of the electric light in remote stations and in countries where the labour available is not capable of much training.

It may possibly be said that in this Paper no definite conclusions are reached as to whether electricity or some other agent is the best source of light in lighthouses generally, nor yet, if electricity be adopted, what is the best way of producing the light and optically dealing with it. The answer is that it is impossible on many points to arrive at general conclusions. Each case must be judged according to its special circumstances.

Extract from Dr Hopkinson's Reply to the discussion on the Paper:—

Professor Adams had touched upon one suggestion of interest, that in order to obtain, with the exceedingly small centre of light afforded by the electric arc, that necessary divergence ordinarily given by the magnitude of the flame, the arc should be placed somewhat out of focus, that the lens should be approximated somewhat towards the centre of light. No doubt if engineers were under the obligation to make use of precisely the same apparatus for electric light as for oil light, the suggestion would be a desirable one to adopt; but it was far better, in the case of electric light, to make the optical apparatus suit that light, and not to suit something else. Even although the apparatus were already in existence for which it was proposed to use the electric arc instead of the oil flame, it would be better economy to remove that apparatus and employ it in some other situation, and to design a special apparatus for the electric light.

Mr Kapp had dealt with the question of coupling alternate-current machines parallel without mechanical connection. In the arrangement at Tino two machines were driven from one counter-shaft. In the first instance, when it was attempted to drive the machines together, there was a difference in the size of the pulleys of about 1 millimetre in 300 millimetres. Notwithstanding that, they synchronized perfectly, and the effect was sufficient to control the natural difference in the time of rotation of the machines. He was not prepared off-hand to say how far it was possible to go in that direction.

3.

ON ELECTRIC LIGHTING.

[From the *Proceedings of the Institution of Mechanical Engineers*, April 25, 1870.]

(First Paper.)

DURING the last year much has been written and much information communicated concerning the production of light from mechanical power by means of an electric current. The major portion of what has appeared has been either descriptive of particular machines for producing the current, and of lamps for manifesting a portion of its energy as light; or a statement of practical results connecting the light obtained with the power applied and the money expended in producing it.

Whilst fully appreciating the present value of such information, the author has felt that it did not tell all that was interesting or practically useful to know. It is desirable to know what the various machines can do with varied and known resistances in the circuit, and with varied speeds of rotation: and what amount of power is absorbed in each case. It is a question of interest, whether a machine intended for one light can or cannot produce two in the same circuit, and if not, why not; whether a machine, such as the Wallace-Farmer, intended as it is for many lights, will give economical results when used for one; and so on. It is clear that the attempt to examine all separate combinations of so many variables would be hopeless, and that the work must be systematised.

The mechanical energy communicated by the steam-engine or other motor is not immediately converted into the energy of heat, but is first converted into the energy of an electric current in a conducting circuit; of this a portion only becomes localised as heat between the carbons of the electric arc; and of this again a part only becomes sensible to the eye as light. The whole of what we need to know may be more easily ascertained and more shortly expressed if the enquiry is divided into two parts: (*a*) what current will a machine produce under various conditions of circuit, and at what expenditure of mechanical power; (*b*) having given the electric conditions under which the arc is placed, no matter how these conditions are produced, what light will be obtained therefrom. Parts of the subject have been treated more or less in this sense by Edlund (*Pogg. Annal.*, 1867 and 1868), Houston and Thomson in America, Mascart (*Journal de Physique*, March 1878), Abney (*Proceedings Royal Society*, 1878), Trowbridge (*Philosophical Magazine*, March 1879), Schwendler (*Report on Electric Light Experiments*), &c. but not so completely that nothing remains to be done; nor does the author doubt that a great deal of information is in the hands of makers of machines, which they have not thought it necessary to make known. The present communication is limited to an account of some experiments on the production of currents by a Siemens medium-sized machine; that is, the machine which is advertised to produce a light of 6000 candles by an expenditure of $3\frac{1}{2}$ horse-power.

All the machines for converting mechanical power into an electric current consist ultimately of a conducting wire moving in a magnetic field; and approximately the electromotive force of the machine will be proportional to the velocity with which the circuit moves through the field, and to the intensity of the field. In general the intensity of the field is not constant; and in such machines as the Siemens and the ordinary Gramme machine it may be regarded as a function of the current passing. We must learn what this function is for the machine in question; or—which comes to exactly the same thing, and is better so long as the facts are merely the result of experiment—we must construct a curve in which the abscissæ represent the intensities of currents passing, and the ordinates the corresponding electromotive forces for a given speed of rotation. But the power of a current, that is its energy per second, is the product of the electromotive force and the intensity, or, in the case of the curve, the product of the ordinate and the abscissa; this is in all cases less than the power required to drive the machine, and the ratio between the two may fairly be called the efficiency of the machine.

The object of the enquiry may perhaps be made clearer by an illustration. Consider the case of a pump forcing water through a pipe against friction; then the electric current corresponds to the volume of water passing per second, and the electromotive force to the difference of pressure on the two sides of the pump; and just as the product of pressure and volume per second is power, so the product of electromotive force and current is power, which is directly comparable with the power expended in driving the machine or the pump, as the case may be. The peculiarity of the so-called dynamo-electric machine lies in this, that what corresponds to the difference of pressure (the electromotive force) depends directly on what corresponds to the volume passing (the current).

Each experiment requires the determination of the speed, the driving power, the resistances in circuit, and the current passing: or of the difference of potential between the two ends of a known resistance of the circuit.

The apparatus employed by the author was arranged, not alone with an aim to accuracy, but in part to make use of such instruments as he happened to possess or could easily construct, and in part with a view to ready erection and transport. Much more accurate results may be obtained by anyone who will arrange apparatus with a single aim to attain the greatest accuracy possible. The author's apparatus will however be briefly described, that others may form their own opinion of the importance of the various sources of error.

The speed-counter was that supplied with the electric machine.

Concerning the steam-engine nothing need be said, save that its speed was maintained very constant by means of a governor, shown in Fig. 1, specially arranged for great sensibility. By placing the joint A above the joint B, instead of below it as in Porter's governor, any degree of sensibility up to instability may be obtained. The speed was varied by means of a weight and a spring, attached to a lever on the throttle-valve spindle. The ungainly appearance of this governor could easily be remedied by anyone proposing to manufacture it.

Governor.

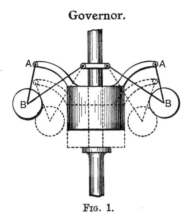

Fig. 1.

The power is transmitted from the engine to a countershaft by means of a strap, and by a second strap from the countershaft to the pulley of the electric machine. On this second strap is the dynamometer shown in Fig. 2.

This dynamometer has for some time been used by Messrs Siemens, and was also used by Mr Schwendler; its invention is due to Herr von Hefner-Alteneck. A is the driving pulley; B the pulley of the electric machine; CC are a pair of loose pulleys between which the strap passes; these are carried in a double triangular frame, which can turn about a bar D. This bar might form part of a permanent structure; but in order to place the dynamometer readily on any strap, the bar was in this case provided with eyes at either end, and secured in position by six or eight ropes. This plan answers well, as there is very little stress on the bar. Immediately above the pulleys CC a cord leads from the frame through a Salter spring-balance over snatch-blocks to a back balance-weight; the tension of this cord is read on the spring balance. At first the spring balance was omitted, and the weight at the end of the cord was observed; but the friction of the snatch-block pulleys was found objectionable. The pulley frame carries a pointer, which is adjusted so as to coincide with a datum mark when the line AB bisects the distance between the loose pulleys. Let W' be the tension of the cord required to bring the pulley frame to its standard position when no work is being

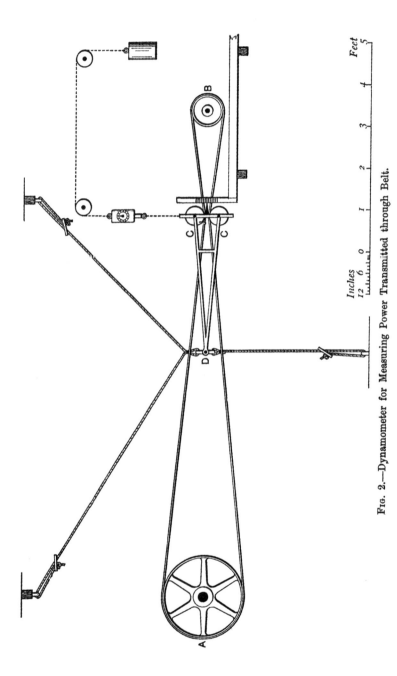

Fig. 2.—Dynamometer for Measuring Power Transmitted through Belt.

transmitted; W'' the tension which is required to bring the pointer back to the datum mark, when an observation is made; and let $W = W' - W''$. Let T', T'' be the tensions on the tight and slack halves of the strap; R_1, R_2, r the radii of the pulleys A, B, and C, plus half the thickness of the strap; c_1, c_2 the distances AJ, JB; $2d$ the distance apart of the centres CC; α_1, α_2 the inclinations of the two parts of the strap, on either side of CC, to the line AB. Then

$$(T' - T'')(\sin \alpha_1 + \sin \alpha_2) = W;$$

$$\text{and } \sin \alpha_1 = \frac{R_1 + r - d}{c_1} + \frac{d}{2c_1}\left(\frac{R_1 + r - d}{c_1}\right)^2,$$

$$\sin \alpha_2 = \frac{R_2 + r - d}{c_2} + \frac{d}{2c_2}\left(\frac{R_2 + r - d}{c_2}\right)^2,$$

very nearly.

The value of $T' - T''$ and the velocity of rotation of the electric machine being known, the power received by it is readily obtained, expressed in gram-centimetres per second. Multiplying by 981, the value of gravity in centimetres and seconds, the power is then expressed in ergs* per second, and is ready for comparison with the results of the electrical experiments.

As already stated, the dynamo-electric machine in the present case was a Siemens medium size; the armature coil has fifty-six divisions; and the brushes are single, not divided, that is, each brush is in connection with one segment of the commutator at any instant.

The leading wire is 100 yards of Siemens No. 90, consisting of seven copper wires, insulated with tape and india-rubber, and having a diameter of about $9\frac{1}{2}$ mm.

The method of determining the current is shown in the diagram, Fig. 3, Plate 27. The current is conveyed from the machine A, through a set of coils of brass wire c, and in some cases through a resistance coil placed in a calorimeter B, and so back to the machine, the connections being made through cups of mercury

---

* The dyne is the force which will in one second impart to one gram a velocity of one centimetre per second, and an erg is the work done by a dyne working through a centimetre; a horse-power may be taken as three-quarters of an erg-ten per second, an erg-ten being 10^{10} ergs. See *Report of Brit. Assoc.*, 1873; and Everett "On the C.G.S. System of Units," published by the Physical Society.

38 ON ELECTRIC LIGHTING.

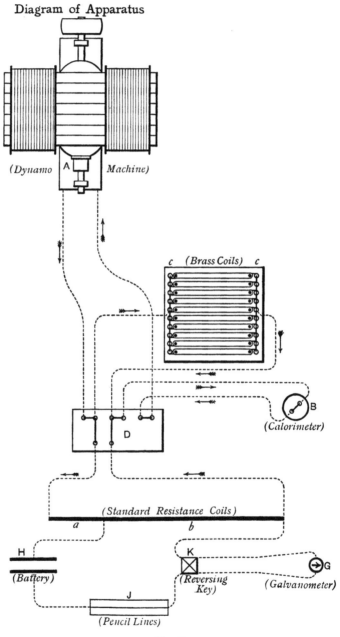

Fig. 3.

excavated in a piece of wood D. The current passing may be ascertained by the heating of the calorimeter, or by measuring the difference of potential at the extremities of the resistance c, all the resistances of the circuit being supposed known. This difference of potential could of course be very easily measured by means of a quadrant electrometer; but, as the instrument had to be frequently removed, a galvanometer appeared more convenient. The two points to be measured are connected to the ends of two series of resistance coils a, b. The galvanometer G is placed in a second derived current, passing from a junction in ab through a battery H, then through a set of high resistances J for adjusting sensibility, a reversing key K, the galvanometer G, the reversing key K again, and so to the other extremity of b. The electromotive force is ascertained by adjusting the resistance b so that the deflection of the galvanometer is nil.

The resistance coils c comprise ten coils of common brass wire, each wound round a couple of wooden uprights driven into a baseboard common to the set; each wire is about 60 metres long, and of No. 17 Birmingham wire-gauge (·06 inch or $1\frac{1}{2}$ mm. diameter), weighing about 14·6 grams per metre. Each terminal is connected to a cup of mercury excavated in the baseboard, so that the coils can be placed in series or in parallel circuit at pleasure. The resistance of each coil being about 3 ohms, this set may be arranged to give resistances varying from 0·3 to 30 ohms.

The calorimeter B is a Siemens pyrometer with the top scale removed; a resistance coil of uncovered German-silver wire nearly 2 m. long, $1\frac{1}{2}$ mm. in diameter, and having a resistance of about 0·2 ohm, is suspended within it from an ebonite cover, which also carries a little brass stirrer; and the calorimeter is filled with water to a level determined by the mark of a scriber. It was of course necessary to know the capacity of the calorimeter for heat. It was filled with warm water up to the mark, and the coil placed in position; 120 grams of water were then withdrawn, and the temperature of the calorimeter was observed to be 58·8° C.; after the lapse of one minute it was 58·3° C.; after a second minute 57·9° C.; 120 grams of cold water, temperature 13·3° C., were then suddenly introduced through a hole in the ebonite cover, and it was found that, two minutes after the reading of 57·9° C., the temperature was 50·0° C.; hence we may infer that the capacity of the calorimeter is equal to that of 740 grams of water. Two similar experiments at lower temperatures gave respectively the numbers 749 and 750. Estimating the capacity from the weight of the copper cylinder supplied with the pyrometer, it should be 747, to which must be added the capacity of the German-silver wire and stirrer. Taking everything into consideration, 750 grams may be assumed as the most probable result.

The resistance coils a, b are of German-silver, made by Messrs Elliott Brothers; they are on the binary scale from $\frac{1}{2}$ ohm to 1024 ohms. Separate coils were used, instead of a regular resistance box, because they were more readily applicable to any other purpose for which they might be required; and the binary scale was adopted, because the coils could at once be used as conductivity coils in parallel circuit, also on the binary scale. Each coil as supplied terminated in two stout copper legs; these were fitted with cups of india-rubber tubing for mercury, whereby any connections whatever could readily be made. This arrangement though rude was very convenient, and perhaps even safer from error than a box with brass plugs to make the connections. By a slight alteration of the connections the whole was instantly available as a Wheatstone's bridge to determine resistances.

The battery H is a single element of Daniell's battery, in which the sulphate of zinc solution floats on the sulphate of copper; its electromotive force is assumed to be $\frac{9}{8}$ volt.

The resistances J added in the battery circuit are pencil lines on glass, such as are described in the *Philosophical Magazine* of February 1879. Three were used, giving a range of sensibility approximately in the proportions 1, 25, 170, 700—the last figure being when all were short-circuited; they are very useful in adjusting the resistance b so as to give no deflection of the galvanometer.

The reversing key K belongs to Sir W. Thomson's electrometer, and is quite suitable when high resistances and nil methods are used.

The galvanometer G is far more sensitive than necessary, and has a resistance of 7000 ohms.

Preliminary to experiments on the current, determinations of resistances were made. The resistance of each brass coil c was first determined, to afford the means of calculating the value of this resistance in any subsequent experiment. When the ten coils are coupled in parallel circuit, the calculated resistance was 0·29 ohm, whilst 0·292 was obtained by direct measurement. The leading wire was then examined; the further ends being disconnected, the insulation resistance was found to be over 60,000 ohms; how much over, it was immaterial to learn. When the ends of the wire were connected, the resistance was found to be 0·129 ohm. The resistances in the dynamo-electric machine A were found to be as follows when cold: magnet coils 0·156 and 0·152 respectively, armature coil 0·324; total 0·632 ohm. Direct examination was made of the whole machine in eight positions of the commutator, giving 0·643 ohm, with a maximum variation of 0·6 per cent. from the mean. After running the machine for some time the resistance was found to be 0·683, an increase which would be accounted for by a rise of temperature of 12° C. or thereabouts. The resistance of the calorimeter B is 0·20 ohm, without its leading wire, which may be taken as 0·01. We have then in circuit three resistances which must be considered: (1) the resistance of the machine A and leading wire, assumed throughout as together 0·81 ohm and denoted by c_1; (2) the resistance of the brass coils c, calculated from the several determinations, with the addition of 0·02 ohm, the resistance of the leading wire, and denoted by c_2; (3) when present, the resistance of the calorimeter B and leading wire, denoted by c_3.

Two approximate corrections were employed, and should be detailed. The first is the correction for the considerable heating of the resistance coils c. These were arranged in two sets of five each, five being in parallel circuit, and the two sets in series. The current from the machine, being about 7·4 webers in each wire, was passed for three or four minutes; the circuit was then broken, and the resistance c_2 was determined within one second of breaking circuit, when it was found to be about 5 per cent. greater than when cold. As the resistance was falling, the following was adopted as a rule of correction: square the current in a single wire, and increase the resistance c_2 by $\frac{1}{10}$th per cent. for every unit in the square. The second correction is due to the fact that the calorimeter was losing heat all the time it was being used. It was assumed that it loses 0·01° C. per minute for every 1° C. by which the temperature of the calorimeter exceeds that of the air; this correction is of course based on the experiment already mentioned.

The method of calculation may now be explained:—

R is the total resistance of the circuit in ohms, equal to $c_1 + c_2 + c_3$;

Q is the current passing in webers;

E is the electromotive force round the circuit in volts;

W_1 is the work per second converted into heat in the circuit, as determined by the galvanometer, measured in erg-tens per second;

W_2 is the work per second as determined by the calorimeter;

W_3 is the work per second as determined by the dynamometer, less the power required to drive the machine when the circuit is open;

HP is the equivalent of W_3 in horse-power;

n is the number of revolutions per minute of the armature.

As already mentioned, the standard resistance coils a, b are adjusted in each experiment so that the galvanometer gives no deflection, and the value of b is then noted. The values of c_1, c_2, c_3 are known from previous observations. Then

$$Q = \tfrac{9}{8} \times \frac{a+b}{b} \times \frac{1}{c_2},$$

$$E = Q \times R,$$

$$W_1 = E \times Q,$$

$$W_2 = \frac{R}{0 \cdot 2} \times \left\{ \begin{matrix} \text{mechanical equivalent of heat} \\ \text{generated per second in calorimeter.} \end{matrix} \right\}$$

The results of the experiments are given in the accompanying Table.

A power of 0·21 erg-ten, or 0·28 horse-power, was required to drive the machine at 720 revolutions on open circuit. An examination of the Table shows that the efficiency of the machine is about 90 per cent., exclusive of friction. Comparing experiments 11 and 13, and also the last four experiments, it is seen that the electromotive force is proportional to the speed of rotation within the errors of observation. Experiments 14, 15, and 16 were

TABLE I.—EXPERIMENTS ON SIEMENS DYNAMO-ELECTRIC MACHINE.

No. of Experiment.	Total Resistance of Circuit. R	Electric Current. Q	Electromotive Force. E	Work measured per second. By Galvanometer. W_1	By Calorimeter. W_2	By Dynamometer. W_3	By Dynamometer. Horse-power.	Revs. of Armature per minute. n	Position of Commutator Brush.
	Ohms.	Webers.	Volts.	Erg-tens.	Erg-tens.	Erg-tens.	H.-P.	Revs.	
1	1025	0·0027	2·72	......	......	......	......	720	Brush in original position, as supplied from maker.
2	8·3	0·48	3·95	0·0019	......	......	......	,,	
3	5·33	1·45	7·73	0·0112	......	0·042	0·056	,,	
4	4·07	16·8	68·4	1·149	1·140	1·179	1·59	,,	
5	3·88	18·2	70·6	1·285	......	1·263	1·68	,,	
6	3·205	24·8	79·5	1·972	2·158	2·106	2·81	,,	
7	3·025	26·8	81·1	2·174	......	2·392	3·19	,,	
8	2·62	32·2	84·4	2·718	2·888	2·780	3·71	,,	
9	2·43	34·5	83·8	2·894	......	3·370	4·49	,,	
10	2·28	37·1	84·6	3·138	2·903	3·538	4·72	,,	
11	2·08	42·0	87·4	3·671	......	3·960	5·28	,,	
12	1·345	64·0	86·1	5·510	5·349	5·777	7·70	654	(Strap slipping.)
13	2·08	41·1	85·5	3·514	......	3·790	5·05	698	
14	2·07	36·0	74·5	2·682	......	2·852	3·80	696	13° extra lead of brush.
15	2·09	42·7	89·2	3·809	......	4·233	5·64	708	8° ⎫
16	2·09	41·7	87·2	3·636	......	4·135	5·51	713	5° ⎬ less lead of brush than in original position.
17	2·10	45·0	94·5	4·252	......	4·810	6·41	759	5° ⎪
18	2·08	40·8	84·9	3·464	......	4·010	5·35	696	5° ⎪
19	2·06	35·1	72·3	2·538	......	2·672	3·56	586	5° ⎭

intended to ascertain the effect of displacing the commutator brushes.

The principal object of the experiments was to ascertain how the electromotive force depended on the current. This relation is represented by the curve shown in Fig. 4, in which the abscissæ represent the currents flowing, or the values of Q in the

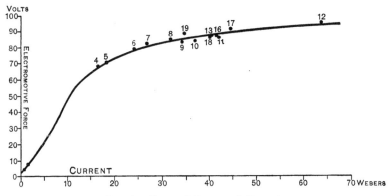

FIG. 4.—Curve of Force and Current.

table, and the ordinates the electromotive forces, or the values of E reduced to a speed of 720 revolutions per minute. The curve may also be taken to represent the intensity of the magnetic field. It will be remarked that there is a point of inflexion in the curve somewhere near the origin. The experiments 1 to 5 indicate that this is the true form of the curve, and it is confirmed in a remarkable manner by a special experiment. A resistance intermediate between $5\frac{1}{3}$ and 4 (experiments 3 and 4) was used in circuit, and E and Q were determined in two different ways: first, by starting with an open circuit, which was then closed; secondly, by starting with a portion of the resistance short-circuited, and a very powerful current passing, and then breaking the short circuit. It was found that E and Q were four times as great in the latter case as in the former. Unfortunately the numbers are not sufficiently accurate to be given, as the solutions of the standard battery had become mixed.

The curve really gives a great deal more information than appears at first sight. It will determine what current will flow at any given speed of rotation of the machine, and under any conditions of the circuit, whether of resistances or of opposed electromotive forces. It will also give very approximate indications of the corresponding curve for other machines of the same configuration, but in which the number of times the wire passes round the electromagnet or the armature is different.

It will be well to compare these results with those obtained by others. M. Mascart worked on a Gramme machine with comparatively low currents: he represents his results approximately by the formula
$$E = n(a + bQ),$$
where a and b are constants. This corresponds to the rapidly rising part of the curve in Fig. 4. Mr Trowbridge with a Siemens machine obtained a maximum efficiency of 76 per cent., and states that the machine was running below its normal velocity. Mr Schwendler's results, when fully published, will probably be found to be the most complete and most accurate existing. In the *précis* he states that the loss of power with a Siemens machine in producing currents of over 20 webers is 12 per cent. Now taking the author's experiments 4 to 19, the mean value of W_1 is 3·027 erg-tens and of W_3 3·304: adding to the latter 0·21, the power required to drive the machine when no current passes, it appears that 13·8 per cent. of the power applied is wasted. Again, taking experiments 4, 6, 8, 10, and 12, the mean value of W_2 is 2·888 erg-tens and of W_3 3·076, indicating a waste of power amounting to 12 per cent. Of this, as already stated, 0·21 erg-ten or 0·28 horse-power is accounted for by friction of the journals and commutator brush; the remainder is expended in local currents, or by loss of kinetic energy of current when sparks occur at the commutator.

According to Weber's theory of induced magnetism, as set forth in Maxwell's *Electricity*, vol. II., if X be the magnetising force and I the intensity of magnetisation,
$$I = \tfrac{2}{3}a\frac{X}{b}, \text{ until } X \text{ rises to the value } b,$$
$$\text{and } I = a\left(1 - \tfrac{1}{3}\frac{b^2}{X^2}\right), \text{ if } X > b,$$
where a and b are constants. We should naturally expect that a similar formula would be approximately applicable to dynamo-electric machines.

In the present experiments, let I be the electromotive force, X the current passing, and assume a to be 60 and b to be 15; we then obtain results not far from those of experiment. The capacity of any continuous-current machine may thus be shortly stated by giving the values of a and b; or, which comes to the same thing, by stating the electromotive force at a given speed when the current is as great as possible, and also the total resistance through which the machine will exert an electromotive force two-thirds of this greatest electromotive force. To this should be added a statement of the resistance of the machine, and of the power it absorbs, with known conditions of the circuit.

The author has not yet tried any quantitative experiments with the electric light, but hopes shortly to do so. In the meantime he would remark that, as the lamp is usually adjusted, only half the energy of the current appears in the arc, or 44 per cent. of the energy transmitted to the machine by the strap.

In conclusion the author would express the obligation he is under to Messrs Chance Brothers and Co., on account of the facilities he has enjoyed for making these experiments at their works. It may be mentioned that one principal object of the research of which this is a beginning is to obtain a minute knowledge of the electric light, with a view to lighthouse illumination.

Dr J. HOPKINSON said that since the curve, Fig. 4, Plate 29, was drawn out, it had occurred to him that the curious circumstance of the curve starting in the manner indicated and then bending upwards from the point of inflection might be accounted for in another way. It might be supposed that the magnetisation of the iron in the machine lagged a little behind the current, so that, if the current were decreasing, the magnetisation for the same current would be greater than if it were increasing. If that were so, the curve would be different as the current was rising or falling; and thus, instead of a single curve with a point of inflection, there would be two curves at a little distance from each other. He also thought a sufficiently accurate formula for the curve in question might probably be based upon Weber's theory of induced magnetism in iron, as he had suggested at the end of the paper. It would be a formula with only two constants in it, so that a full description could be given briefly of all the curve had to tell.

4.

ON ELECTRIC LIGHTING.

[From the *Proceedings of the Institution of Mechanical Engineers*, pp. 266—274. April 23, 1880.]

(*Second Paper.*)

Dynamo-electric Machines.—Since the date of the author's former paper in April 1879, other observers have published the results of experiments similar to those described by him. It may be well to exhibit some of these results reduced to the form he has adopted, namely, a curve, such as that previously shown, and now

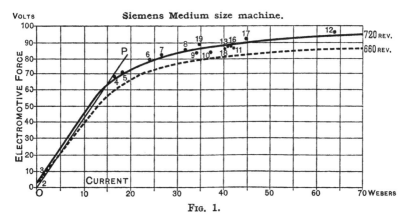

Fig. 1.

reproduced, with slight alterations, in Fig. 1. Here any abscissa represents a current passing through the dynamo-electric machine, and the corresponding ordinate represents the electromotive force of the machine for a certain speed of revolution, when that current is passing through it. It will be found, (1) that with varying speed the ordinate, or electromotive force, corresponding to any abscissa or current, is proportional to the speed; (2) that the electromotive force does not increase indefinitely with increasing current, but that the curve approaches an asymptote; (3) that the earlier part of the curve is, roughly speaking, a straight line, until the current attains a certain value, and that at that point the electromotive force has reached about two-thirds of its maximum value. When the current is such that the electromotive force is not more than two-thirds of its maximum, a very small change in the resistance with speed of engine constant, or in the speed of the engine with resistance constant, causes a great change in the current. For this reason the greatest of these currents, which is that corresponding to the point where the curve breaks away from a straight line, and which is the same for all speeds of revolution, since the curves for different speeds differ only in the scale of ordinates, may be called the "critical current" of the machine. The effect of a change of speed is exhibited in Fig. 1, where the lower dotted line represents the curve for a speed of 660 revolutions per minute, instead of 720. The resistance, varying as $\frac{\text{electromotive force}}{\text{current}}$, is given by the slope of the line OP. But since the resistance is constant, the slope of this line must be constant; and it will be seen that it cuts the upper curve at a point corresponding to a current of 15 webers, and the lower at a point corresponding to a current of 5 webers only.

In Germany, Auerbach and Meyer (*Wiedemann's Annalen*, Nov. 1879) have experimented fully on a Gramme machine at various speeds, and with various external resistances. The resistance of the machine was 0·97 ohm. Their results are summarised in a Table at the end of their paper, which gives the current passing, with resistances in circuit from 1·75 to 200 Siemens units, and at speeds from 20 to 800 revolutions per minute. In the accompanying diagram, Fig. 2, Plate 30, the curve G expresses the relation between electromotive force and current, as deduced from some of their observations; the points marked are plotted from their Table, making allowance, where necessary, for difference in speed. The curve, as actually constructed, is for a speed of 800 revolutions: at this speed it will be seen that the maximum electromotive force is about 76 volts; and the critical current, corresponding to a force of about 51 volts, is 6·5 webers, with a total resistance of 7·8 ohms. Up to this point there will be great instability, exactly as was the case in the Siemens machine examined by the author, where the resistance was 4 ohms, and the speed 720 revolutions.

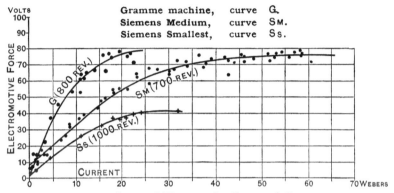

Fig. 2. Curves of Electromotive Force and Current.

The results of an elaborate series of experiments on certain dynamo-electric machines have recently been presented to the Royal Society by Dr Siemens. One of the machines examined was an ordinary medium-sized machine, substantially similar to that tried by the author in 1879. It is described as having 24 divisions of the commutator; 336 coils on the armature, with a resistance of 0·4014 Siemens units; and 512 coils on the magnets, with a resistance of 0·3065; making a total resistance of 0·7079 Siemens units = 0·6654 ohms. The curve Sm, Fig. 2, gives the relation of electromotive force and current, reduced to a speed of 700 rev. per minute, the actual speeds ranging from 450 to 800 rev. The maximum electromotive force appears to be probably 76 volts, and the critical current 15 webers: which is the same as in the author's first experiments on a similar machine.

In the summer of 1879 the author examined a Siemens machine of the smallest size. This machine is generally sold as an exciter for their alternate-current machine. It has an internal resistance of 0·74 ohms, of which 0·395 is in the armature or helix. The machine is marked to run at 1130 rev. per min. The following Table II. gives, for a speed of 1000 rev., the total resistance, current, electromotive force, and horse-power developed as current. The horse-power expended was not determined.

TABLE II.—EXPERIMENTS ON SMALLEST-SIZED SIEMENS DYNAMO-ELECTRIC MACHINE.

Resistance.	Electric Current.	Electromotive Force.	Horse-power developed as current.
Ohms.	Webers.	Volts.	H. P.
2·634	5·10	13·2	0·09
2·221	12·15	27·0	0·44
1·967	17·0	33·6	0·76
1·784	20·4	36·4	0·99
1·668	22·3	37·2	1·11
1·579	23·2	36·6	1·14
1·503	25·6	39·3	1·34
1·440	27·8	40·0	1·49
1·145	36·2	41·5	2·00

The curve Ss, Fig. 2, gives as usual the relations of electromotive force and current. From this curve it will be seen that the critical current is 11·2 webers, and the maximum electromotive force, at the speed of 1000 rev., is about 42 volts. The determinations for this machine were made in exactly the same manner as in the experiments on the medium-sized machine, using the galvanometer, but omitting the experiment with the calorimeter (compare Table I., *Proceedings Inst. Mech. Engineers*, page 249, 1879).

The time required to develop the current in a Gramme machine has been examined by Herwig (Wiedemann, June 1879). He established the following facts for the machine he examined. A reversed current, having an electromotive force of 0·9 Grove cells, sufficed to destroy the residual magnetism of the electro-magnets. If the residual magnetism was as far as possible reduced, it took a much longer time to get up the current than when the machine was in its usual state. A longer time was required to get up the current when the external resistance was great than when it was small. With ordinary resistance the current required from $\frac{3}{4}$ second to 1 second to attain its maximum.

Brightness of the Electric Arc.—The measurement of the light emitted by an electric arc presents certain peculiar difficulties.

The light itself is of a different colour from that of a standard candle, in terms of which it is usual to express luminous intensities. The statement, without qualification, that a certain electric lamp and machine give a light of a specified number of candles, is therefore wanting in definite meaning. A red light cannot with propriety be said to be any particular multiple of a green light; nor can one light, which is a mixture of colours, be said with strictness to be a multiple of another, unless the proportions of the colours in the two cases are the same. Captain Abney (*Proceedings of the Royal Society*, 7 March 1878, p. 157) has given the results of measurements of the red, blue, and actinic light of electric arcs, in terms of the red, blue, and actinic light of a standard candle. The fact that the electric light is a very different mixture of rays from the light of gas or of a candle, has long been known, but has been ignored in statements intended for practical purposes.

Again, the emission of rays from the heated carbons and arc is by no means the same in all directions. Determinations have been made in Paris of the intensity in different directions, in particular cases. If the measurement is made in a horizontal direction, a very small obliquity in the crater of the positive carbon will throw the light much more on one side than on the other, causing great discordance in the results obtained.

If the electric light be compared directly with a standard candle, a dark chamber of great length is needed—a convenience not always attainable. In the experiments made at the South Foreland by Dr Tyndall and Mr Douglass, an intermediate standard was employed; the electric light was measured in terms of a large oil lamp, and this latter was frequently compared with a standard candle.

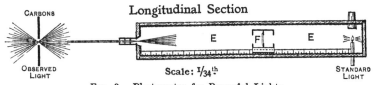

Fig. 3.—Photometer for Powerful Lights.

Other engagements have prevented the author from fairly attacking these difficulties; but since May 1879 he has had in occasional use a photometer with which powerful lights can be measured in moderate space. This photometer is shown in Figs. 3 and 4, and an enlargement of the field-piece in Fig. 5. A convex lens A, of short focus, forms an image at B of the powerful source of light which it is desired to examine. The intensity of the light from this image will be less than that of the actual source by a calculable amount; and when the distance of

Transverse Section

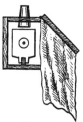

Fig. 4.

the lens from the light is suitable, the reduction is such that the reduced light becomes comparable with a candle or a carcel lamp. Diaphragms CC are arranged in the cell which contains the lens, to cut off stray light. One of these is placed at the focus of the lens, and has a small aperture. It is easy to see that this diaphragm will cut off all light entering from a direction other

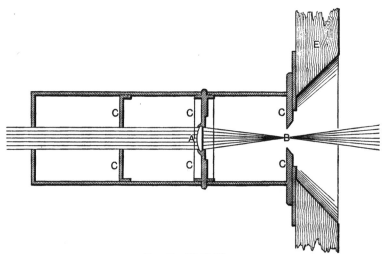

Fig. 5.—Field Piece.

than that of the source; so effectually does it do so, that observations may be made in broad daylight on any source of light, if a dark screen be placed behind it. The long box EE, Fig. 3, of about 7 ft. length, is lined with black velvet,—the old-fashioned dull velvet, not that now sold with a finish, which reflects a great deal of the light incident at a certain angle. This box serves as a dark chamber, in which the intensity of the image formed by the lens is compared with a standard light, by means of an ordinary Bunsen's photometer F, sliding on a graduated bar.

Mr Dallmeyer kindly had the lens made for the author: he can therefore rely upon the accuracy of its curvature and thickness; it is plano-convex, the convex side being towards the source of light. The curvature is exactly 1 in. radius and the thickness is 0·04 in.; it is made of Chance's hard crown-glass, of which the refractive index for the D line in the spectrum is 1·517. The focal length f is therefore 1·933 inch.

Let u denote the distance of the source of light from the curved surface of the lens, and v the distance of the image B of the source from the posterior focal plane. Neglecting for the moment loss by reflection at the surface of the glass, the intensity of the source is reduced by the factor $\left(\frac{v}{u}\right)^2$. But $\frac{1}{v}+\frac{1}{u}=\frac{1}{f}$, or $v=\frac{uf}{u-f}$; hence the factor of reduction is $\left(\frac{f}{u-f}\right)^2$. The effect of absorption in so small a thickness of very pure glass may be neglected; but the reflection at the surfaces will cause a loss of 8·3 per cent., which must be allowed for. This percentage is calculated from Fresnel's formulæ, which are certainly accurate for glasses of moderate refrangibility, and for moderate angles of incidence.

Suppose, for example, it is required to measure a light of 8000 candles; if it be placed at a distance of 40 in., it will be reduced in the ratio 467 to 1, and becomes a conveniently measurable quantity. By transmitting through coloured glasses both the light from an electric lamp and that from the standard, a rough comparison may be made of the red or green in the electric light with the red or green in the standard.

A dispersive photometer, in which a lens is used in a somewhat similar manner, is described in Stevenson's *Lighthouse Illumination;* but in that case the lens is not used in combination with a Bunsen's photometer, nor with any standard light. Messrs Ayrton and Perry described a dispersive photometer with a concave lens at the meeting of the Physical Society on 13th December 1879 (*Proceedings Physical Society*, Vol. III. p. 184). The convex lens possesses however an obvious advantage in having a real focus, at which a diaphragm to cut off stray light may be placed.

Efficiency of the Electric Arc.—To define the electrical condition of an electric arc, two quantities must be stated—the current passing, and the difference of electric potential at the ends of the two carbons. Instead of either one of these, we may, if we please, state the ratio $\frac{\text{difference of potential}}{\text{current}}$, and call it the resistance of the arc, that is to say, the resistance which would replace the arc without changing the current. But such a use of the term electric resistance is unscientific; for Ohm's law, on which the definition of electric resistance rests, is quite untrue of the electric arc, while on the other hand, for a given material of the electrodes, a given distance between them, and a given atmospheric pressure, the difference of potential on the two sides of the arc is approximately constant. The product of the difference of potential and the current is of course equal to the work developed in the arc; and this, divided by the work expended in driving the machine, may be considered as the efficiency of the whole combination. It is a very easy matter to measure these quantities. The difference of potential on the two sides of the arc may be measured by the method given by the author in his previous paper, or by an electrometer, or in other ways. The current may be measured by an Obach's galvanometer, or by a suitable electro-dynamometer, or best of all, in the author's opinion, by passing the whole current, on its way to the arc, through a very small known resistance, which may be regarded as a shunt for a galvanometer of very high resistance, or to the circuit of which a very high resistance has been added.

It appears that with the ordinary carbons, and at ordinary atmospheric pressure, no arc can exist with a less difference of potential than about 20 volts; and that in ordinary work, with an arc about $\frac{1}{4}$ in. long, the difference of potential is from 30 to 50 volts. Assuming the former result, about 20 volts, for the difference of potential, the use of the curve of electromotive forces may be illustrated by determining the lowest speed at which a given machine can run, and yet be capable of producing a short arc. Taking O as the origin of co-ordinates, Fig. 6, set off upon the axis of ordinates the distance OA equal to 20 volts;

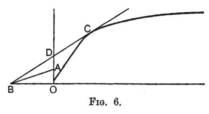

Fig. 6.

draw AB to intersect at B the negative prolongation of the axis of abscissæ, so that the ratio $\dfrac{OA}{OB}$ may represent the necessary metallic resistance of the circuit. Through the point B, thus obtained, draw a tangent to the curve, touching it at C, and cutting OA in D. Then the speed of the machine, corresponding to the particular curve employed, must be diminished in the ratio $\dfrac{OD}{OA}$, in order that an exceedingly small arc may be just possible.

The curve may also be employed to put into a somewhat different form the explanation given by Dr Siemens at the Royal Society, respecting the occasional instability of the electric light, as produced by ordinary dynamo-electric machines. The operation of all ordinary regulators is to part the carbons when the current is greater than a certain amount, and to close them when it is less; initially the carbons are in contact. Through the origin O, Fig. 7, draw the straight line OA, inclined at the angle

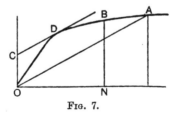

Fig. 7.

representing the resistances of the circuit other than the arc, and meeting the curve at A. The abscissa of the point A represents the current which will pass if the lamp be prevented from operating. Let ON represent the current to which the lamp is adjusted; then if the abscissa of A be greater than ON, the carbons will part. Through N draw the ordinate BN, meeting the curve in the point B; and parallel to OA draw a tangent CD, touching the curve at D. If the point B is to the right of D, or further from the origin, the arc will persist; but if B is to the left of D, or nearer to the origin, the carbons will go on parting, till the current suddenly fails and the light goes out. If B, although to the right of D, is very near to it, a very small reduction in the speed of the machine will suffice to extinguish the light. Dr Siemens gives greater stability to the light by exciting the electro-magnets of the machine by a shunt circuit, instead of by the whole current.

The success of burning more than one regulating lamp in series depends on the use in the regulator of an electro-magnet, excited by a high-resistance wire connecting the two opposed carbons. The force of this magnet will depend upon the difference of potential in the arc, instead of depending, as in the ordinary lamp, upon the current passing. Such a shunt magnet has been employed in a variety of ways. The author has arranged it as an attachment to an ordinary regulator; the shunt magnet actuates a key, which short-circuits the magnet of the lamp when the carbons are too far parted, and so causes them to close.

In conclusion the author ventures to remind engineers of the following rule for determining the efficiency of any system of electric lighting in which the electric arc is used, the arc being neither exceptionally long nor exceptionally short. Measure the difference of potential of the arc, and also the current passing through it, in volts and webers respectively; then the product of these quantities, divided by 746, is the horse-power developed in that arc. It is then known that the difference between the horse-power developed in the arc and the horse-power expended to drive the machine must be absolutely wasted, and has been expended in heating either the iron of the machine or the copper conducting wires.

5.

SOME POINTS IN ELECTRIC LIGHTING.

[*A Lecture delivered at the Institution of Civil Engineers,*
5 *April*, 1883.]

ARTIFICIAL light is generally produced by raising some body to a high temperature. If the temperature of a body be greater than that of surrounding bodies it parts with some of its energy in the form of radiation. Whilst the temperature is low these radiations are not of a kind to which the eye is sensitive; they are exclusively radiations less refrangible and of greater wave-length than red light, and may be called infra red. As the temperature is increased the infra red radiations increase, but presently there are added radiations which the eye perceives as red light. As the temperature is further increased, the red light increases and yellow, green and blue rays are successively thrown off in addition. On pushing the temperature to a still higher point, radiations of a wave-length, shorter even than violet light, are produced, to which the eye is insensitive, but which act strongly on certain chemical substances; these may be called ultra violet rays. It is thus seen that a very hot body in general throws out rays of various wave-lengths, our eyes, it so happens, being only sensitive to certain of these, viz., those not very long and not very short, and that the hotter the body the more of every kind of radiation will it throw out; but the proportion of short waves to long waves becomes vastly greater as the temperature is increased. The problem of the artificial production of light with economy of energy is the same as that of raising some body to such a temperature that it shall give as large a proportion as possible of those rays which the eye happens to be capable of feeling. For practical purposes this temperature is the highest temperature we can produce. Owing to the high temperature at which it remains solid, and to its great emissive power the radiant body used for artificial illumination is nearly always some form of carbon. In the electric current we have an agent whereby we can convert more energy of other forms into heat in a small space than in any other way; and fortunately carbon is a conductor of electricity as well as a very refractory substance.

The science of lighting by electricity very naturally divides itself into two principal parts—the methods of production of electric currents, and of conversion of the energy of those currents into heat at such a temperature as to be given off in radiations to which our eyes are sensible. There are other subordinate branches of the subject, such as the consideration of the conductors through which the electric energy is transmitted, and the measurement of the quantity of electricity passing and its potential or electric pressure. Although I shall have a word or two to say on the other branches of the subject, I propose to occupy most of the time at my disposal this evening with certain points concerning the conversion of mechanical energy into electrical energy. We know nothing as to what electricity is, and its appeals to our senses are in general less direct than those of the mechanical phenomena of matter. The laws, however, which we know to connect together those phenomena which we call electrical, are essentially mechanical in form, are closely correlated with mechanical laws, and may be most aptly illustrated by mechanical analogues. For example, the terms "potential," "current," and "resistance," with which we are becoming familiar in electricity have close analogues respectively in "head," "rate of flow," and "coefficient of friction" in the hydraulic transmission of power. Exactly as in hydraulics head multiplied by velocity of flow is power measured in foot-pounds per second or in horse-power, so potential multiplied by current is power and is measurable in the same units. The horse-power not being a convenient electrical unit Dr Siemens has suggested that the electrical unit of power or volt-ampère should be called a watt: 746 watts are equal to one horse-power. Again, just as water flowing in a pipe has inertia and requires an expenditure of work to set it in motion, and is capable of producing disruptive effects if its motion is too suddenly arrested—as, for example, when a plug-tap is suddenly closed in a pipe through which water is flowing rapidly, so a current of electricity in a wire has inertia; to set it moving electromotive force must work for a finite time, and if we attempt to arrest it suddenly by breaking the circuit the electricity forces its way across the interval as a spark. Corresponding to mass and moments of inertia in mechanics we have in electricity coefficients of self-induction. We will now show that an electric circuit behaves as though it had inertia. The apparatus we shall use is shown diagrammatically in Fig. 1.

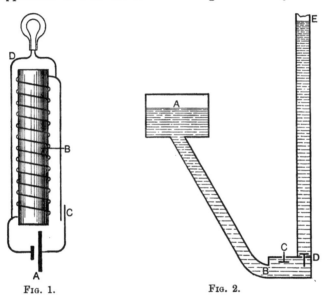

Fig. 1.   Fig. 2.

A current from a Sellon battery A circulates round an electromagnet B; it can be made and broken at pleasure at C. Connected to the two extremities of the wire on the magnet is a small incandescent lamp D, lent to me by Mr Crompton, of many times the resistance of the coil. On breaking the circuit, the current in the coil, in virtue of its momentum, forces its way through the lamp, and renders it momentarily incandescent, although all connection with the battery, which in any case would be too feeble to send sufficient current through the lamp, has ceased. Let us try the experiment, make contact, break contact. You observe the lamp lights up. Compare with the diagram Fig. 2 of the hydraulic analogue the hydraulic ram. There a current of water suddenly arrested forces a way for a portion of its quantity to a greater height than that from which it fell. *AB* corresponds to the electro-magnet, the valve *C* to the contact-breaker, and *DE* to the lamp. There is, however, this difference between the inertia of water in a pipe and the inertia of an electric current—the inertia of the water is confined to the water, whereas the inertia of the electric current resides in the surrounding medium. Hence arise the phenomena of induction of currents upon currents, and of magnets upon moving conductors—phenomena which have no immediate analogues in hydraulics. There is thus little difficulty to anyone accustomed to the laws of rational mechanics in adapting the expression of those laws to fit electrical phenomena; indeed we may go so far as to say that the part of electrical science with which we have to deal this evening is essentially a branch of mechanics, and as such I shall endeavour to treat it.

This is neither the time nor the place for setting forth the fundamental laws of electricity, but I cannot forbear from showing you a mechanical illustration, or set of mechanical illustrations, of the laws of electrical induction, first discovered by Faraday. I have here a model, Fig. 3, which was made to the instructions of the late Professor Clerk Maxwell, to illustrate the laws of induction. It consists of a pulley *P*, which I now turn with my hand, and which represents one electric circuit, its motion the current therein.

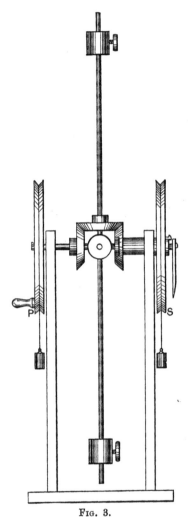

Fig. 3.

Here is a second pulley S, representing a second electric circuit. These two pulleys are geared together by a simple differential train, such as is sometimes used for a dynamometer. The intermediate wheel of the train, however, is attached to a balanced fly-wheel, the moment of inertia of which can be varied by moving inwards or outwards these four brass weights. The resistances of the two electric circuits are represented by the friction on the pulleys of two strings, the tension of which can be varied by tightening these elastic bands. The differential train, with its fly-wheel, represents the medium, whatever it may be, between the two electric conductors. The mechanical properties of this model are of course obvious enough. Although the mathematical equations which represent the relation between one electric conductor and another in its neighbourhood are the same in form as the mathematical equations which represent the mechanical connection between these two pulleys, it must not be assumed that the magnetic mechanism is completely represented by the model. We shall now see how the model illustrates the action of one electric circuit upon another. You know that Faraday discovered that if you have two closed conductors arranged near to and parallel to each other, and if you cause a current of electricity to begin to flow in the first, there will arise a temporary current in the opposite direction in the second. This pulley, marked P on the diagram, represents the primary circuit, and the pulley marked S on the diagram the secondary circuit. We cause a current to begin to flow in the primary, or turn the pulley P; an opposite current is induced in the secondary circuit, or the pulley S turns in the opposite direction to that in which we began to move the pulley P. The effect is only temporary, resistance speedily stops the current in the secondary circuit, or, in the mechanical model, friction the rotation of the pulley S. I now gradually stop the motion of P; the pulley S moves in the direction in which P was previously moving, just as Faraday found that the cessation of the primary current induced in the secondary circuit a current in the same direction as that which had existed in the primary. If there were a large number of convolutions or coils in the secondary circuit, but that circuit were not completed, but had an air space interrupting its continuity, an experiment with the well-known Ruhmkorff coil would show you that when the current was *suddenly* made to cease to flow in the primary circuit, so great an electromotive force would be exerted in the secondary circuit that the electricity would leap across the space as a spark. I will now show you what corresponds to a spark with this mechanical model. The secondary pulley S shall be held by passing a thread several times round it. I gradually produce the current in the primary circuit: I will now suddenly stop this primary current: you observe that the electromotive force is sufficient to break the thread. The inductive effects of one electric circuit upon another depend not alone on the dimensions and form of the two circuits, but on the nature of the material between them. For example, if we had two parallel circular coils, their inductive effects would be very considerably enhanced by introducing a bar of iron in their common axis. We can imitate this effect by moving outwards or inwards these brass weights. In the experiment I have shown you the weights have been some distance from the axis in order to obtain considerable effect, just as in the Ruhmkorff coil an iron core is introduced within the primary circuit. I will now do what is equivalent to removing the core: I will bring the weights nearer to the axis, so that my fly-wheel shall have less moment of inertia. You observe that the inductive effects are very much less marked than they were before. With the same electromagnet which we used before, but differently arranged, we will show what we have just illustrated—the induction of one circuit on another. Referring to Fig. 4, coil AB corresponds to wheel P; CD to wheel S; and the iron core to the fly-wheel and differential gear. The resistance of a lamp takes the place of the friction of the string on S. As we make and break the circuit you see the effect of the induced current in rendering the lamp incandescent. So far I have been illustrating the phenomena of the induction of one current upon another. I will now show on the model that a current in a single electric circuit has momentum. The secondary wheel shall be firmly held; it shall have no conductivity at all; that is, its electrical effect shall be as though it were not there. I now cause a current to begin to flow in the primary circuit, and it is obvious enough that a certain

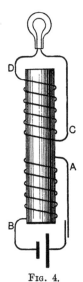

Fig. 4.

amount of work must be done to bring it up to a certain speed. The angular velocity of the fly-wheel is half that of the pulley representing the primary circuit. Now suppose that the two pulleys were connected together in such a way that they must have the same angular velocity in the same direction. This represents the coil having twice as many convolutions as it had before. A little consideration will show that I must do four times as much work to give the primary pulley the same velocity that it attained before; that is to say, that the coefficient of self-induction of a coil of wire is proportional to the square of the number of convolutions. Again, suppose that these two wheels were so geared together that they must always have equal and opposite velocities, you can see that a very small amount of work must be done in order to give the primary wheel the velocity which we gave to it before. Such an arrangement of the model represents an electric circuit, the coefficient of induction of which is exceedingly small, such as the coils that are wound for standard resistances; the wire is there wound double, and the current returns upon itself, as shown in Fig. 5.

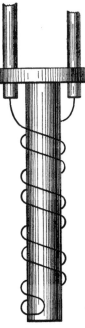

Fig. 5.

In the widest sense, the dynamo-electric machine may be defined as an apparatus for converting mechanical energy into the energy of electro-static charge, or mechanical power into its equivalent electric current through a conductor. Under this definition would be included the electrophorus and all frictional machines; but the term is used, in a more restricted sense, for those machines which produce electric currents by the motion of conductors in a magnetic field, or by the motion of a magnetic field in the neighbourhood of a conductor. The laws on which the action of such machines is based have been the subject of a series of discoveries. Oersted discovered that an electric current in a conductor exerted force upon a magnet; Ampère that two conductors conveying currents generally exerted a mechanical force upon each other: Faraday discovered—what Helmholtz and Thomson subsequently proved to be the necessary consequence of the mechanical reactions between conductors conveying currents and magnets—that if a closed conductor move in a magnetic field, there will be a current induced in that conductor in one direction, if the number of lines of magnetic force passing through the conductor was increased by the movement; in the other direction if diminished. Now all dynamo-electric machines are based upon Faraday's discovery. Not only so; but however elaborate we may wish to make the analysis of the action of a dynamo-machine, Faraday's way of presenting the phenomena of electro-magnetism to the mind is in general our best point of departure. The dynamo-machine, then, essentially consists of a conductor made to move in a magnetic field. This conductor, with the external circuit, forms a closed circuit in which electric currents are induced as the number of lines of magnetic force passing through the closed circuit varies. Since, then, if the current in a closed circuit be in one direction when the number of lines of force is increasing, and in the opposite direction when they are diminishing, it is clear that the current in each part of such circuit which passes through the magnetic field must be alternating in direction, unless indeed the circuit be such that it is continually cutting more and more lines of force, always in the same direction. Since the current in the wire of the machine is alternating, so also must be the current outside the machine, unless something in the nature of a commutator be employed to reverse the connections of the internal wires in which the current is induced, and of the external circuit. We have then broadly two classes of dynamo-electric machines; the simplest, the alternating-current machine, where no commutator is used; and the continuous-current machine, in which a commutator is used to change the connection of the external circuit just at the moment when the direction of the current would change. The mathematical theory of the alternate-current machine is comparatively simple. To fix ideas, I will ask you to think of the alternate-current Siemens machine, which Dr Siemens exhibited here three weeks ago. We have there a series of magnetic fields of alternate polarity, and through these fields we have coils of wire moving; these coils constitute what is called the armature; in them are induced the currents which give a useful effect outside the machine. Now, I am not going to trouble you to go through the mathematical equations, simple though they are, by which the following formulæ are obtained:—

$$I = A \sin \frac{2\pi t}{T} \quad \text{........................(I.)}$$

$$E = \frac{2\pi A}{T} \cos \frac{2\pi t}{T} \quad \text{........................(II.)}$$

$$x = \frac{2\pi A}{T} \frac{\cos 2\pi \frac{t-\tau}{T}}{\sqrt{\left(\frac{2\pi\gamma}{T}\right)^2 + R^2}} \quad \text{............(III.)}$$

$$\tan \frac{2\pi\tau}{T} = \frac{2\pi\gamma}{RT} \quad \text{........................(IV.)}$$

$$\Theta R \frac{2\pi^2 A^2}{T^2} \frac{1}{\left(\frac{2\pi\gamma}{T}\right)^2 + R^2} \quad \text{..................(V.)}$$

T represents the periodic time of the machine; that is, in the case of a Siemens machine having eight magnets on each side of the armature, T represents the time of one-fourth of a revolution. I represents the number of lines of force embraced by the coils of the armature at the time t. I must be a periodic function of t, in the simplest form represented by Equation I. Equation II. gives E the electromotive force acting at time t upon the circuit. Having given the electromotive force acting at any time, it would appear at first sight that we had nothing to do but to divide that electromotive force by the resistance R of the whole circuit, to obtain the current flowing at that time. But if we were to do so we should be landed in error, for the conducting circuit has other properties besides resistance. I pointed out to you that it had a property of momentum represented by its coefficient of self-induction called γ in the formula; and when we are dealing with rapid changes of current, it plays as important a part as the resistance. Formula III. gives the current x, flowing at any time, and you will observe that it shows two things: first, the maximum current is less than it would be if there were no self-induction; secondly, it attains its maximum at a later time. This retardation is represented by the letter τ, and its amount is determined by the Formula IV. At a given speed of rotation, the amount of electrical work developed in the machine in any time Θ is given by Formula V. It is greatest when $R = \frac{2\pi\gamma}{T}$. From these formulæ we see that the current is diminished either by increasing γ or increasing R; also the moment of reversal of current is not coincident with the moment of reversal of electromotive force, but occurs later, by an amount depending on the relative magnitudes of γ and R. They show us that although by doubling the velocity of the machine we really double the electromotive force at any time, we do not double the current passing, nor the work done by the machine; but we may see that if we double the velocity of the machine, we may work through double the external resistance, and still obtain the same current. In what precedes, it has been assumed that the copper wires are the only conducting bodies moving in the magnetic field. In many cases the moving wire-coils of these machines have iron cores, the iron being in some cases solid, in others more or less divided. It is found that if such machines are run on open circuit, that is, so that no current circulates in the armatures; the iron becomes hot, very much hotter than when the circuit of the copper wire is closed. In some cases this phenomenon is so marked that the machine actually takes more to drive it, when the machine is on open circuit, than when it is short-circuited. The explanation is that on open circuit currents are induced in the iron cores, but that when the copper coils are closed the current in them diminishes by induction the current in the iron. The effect of currents in the iron cores is not alone to waste energy and heat the machine; but for a given intensity of field and speed of revolution, the external current produced is diminished. The cure of the evil is to subdivide the moving iron as much as possible, in directions perpendicular to those in which the current tends to circulate.

There remains one point of great practical interest in connection with alternate-current machines, How will they behave when two or more are coupled together, to aid each other in doing the same work? With galvanic batteries, we know very well how to couple them, either in parallel circuit or in series, so that they shall aid, and not oppose, the effects of each other; but with alternate-current machines, independently driven, it is not quite obvious what the result will be, for the polarity of each machine is constantly changing. Will two machines, coupled together, run independently of each other, or will one control the movement of the other in such wise that they settle down to conspire to produce the same effect, or will it be into mutual opposition? It is obvious that a great deal turns upon the answer to this question, for in the general distribution of electric light, it will be desirable to be able to supply the system of conductors from which the consumers draw by separate machines, which can be thrown in and out at pleasure. Now I know it is a common impression that alternate-current machines cannot be worked together, and that it is almost a necessity to have one enormous machine to supply all the consumers drawing from one system of conductors. Let us see how the matter stands. Consider two machines independently driven, so as to have approximately the same periodic time and the same electromotive force. If these two machines are to be worked together they may be connected in one of two ways; they may be in parallel circuit with regard to the external conductor, as shown by the full line in Fig. 6, that is,

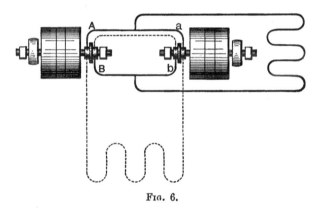

Fig. 6.

their currents may be added algebraically and sent to the external circuit, or they may be coupled in series, as shown by the dotted line, that is, the whole current may pass successively through the two machines, and the electromotive force of the two machines may be added, instead of their currents. The latter case is simpler. Let us consider it first. I am going to show that if you couple two such alternate-current machines in series, they will so control each other's phase as to nullify each other, and that you will get no effect from them; and, as a corollary from that, I am going to show that if you couple them in parallel circuit, they will work perfectly well together, and the currents they produce will be added; in fact, that you cannot drive alternate-current machines tandem, but that you may drive them as a pair, or, indeed, any number abreast. In diagram, Fig. 7, the horizontal line of abscissæ represents the time advancing from left to right; the full curves represent the electromotive forces of the two machines not supposed to be in the same phase. We want to see whether they will tend to get into the same phase or to get into opposite phases. Now, if the machines are coupled in series, the resultant electromotive force

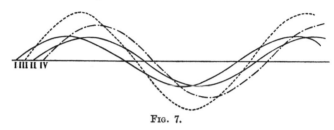

FIG. 7.

on the circuit will be the sum of the electromotive forces of the two machines. This resultant electromotive force is represented by the broken curve III; by what we have already seen in Formula IV, the phase of the current must lag behind the phase of the electromotive force, as is shown in the diagram by curve IV, thus ———.———.———. Now the work done in any machine is represented by the sum of the products of the currents and of the electromotive forces, and it is clear that as the phase of the current is more near to the phase of the lagging machine II than to that of the leading machine I, the lagging machine must do more work in producing electricity than the leading machine; consequently its velocity will be retarded, and its retardation will go on until the two machines settle down into exactly opposite phases, when no current will pass. The moral, therefore, is, do not attempt to couple two independently driven alternate-current machines in series. Now for the corollary, AB, Fig. 6, represent the two terminals of an alternate-current machine; ab the two terminals of another machine independently driven. A and a are connected together, and B and b. So regarded, the two machines are in series, and we have just proved that they will exactly oppose each other's effects, that is, when A is positive, a will be positive also; when A is negative, a is also negative. Now, connecting A and a through the comparatively high resistance of the external circuit with B and b, the current passing through that circuit will not much disturb, if at all, the relations of the two machines. Hence, when A is positive, a will be positive, and when A is negative, a will be negative also; precisely the condition required that the two machines may work together to send a current into the external circuit. You may, therefore, with confidence, attempt to run alternate-current machines in parallel circuit for the purpose of producing any external effect. I might easily show that the same applies to a larger number: hence, there is no more difficulty in feeding a system of conductors from a number of alternate-current machines, than there is in feeding it from a number of continuous-current machines. A little care only is required that the machine shall be thrown in when it has attained something like its proper velocity. A further corollary is that alternate currents with alternate-current machines as motors may theoretically be used for the transmission of power*.

It is easy to see that, by introducing a commutator revolving with the armature, in an alternate-current machine, and so arranged as to reverse the connection between the armature and the external circuit just at the time when the current would reverse, it is possible to obtain a current constant always in direction; but such a current would be far from constant in intensity, and would certainly not accomplish all the results that are obtained in modern continuous-current machines. This irregularity may, however, be reduced to any extent by multiplying the wires of the armature, giving each its own connection to the outer circuit, and so placing them that the electromotive force attains a maximum successively in the several coils. A practically uniform electric current was first commercially produced with the ring armature of Pacinotti, as perfected by Gramme. The Gramme machine is represented diagrammatically in Fig. 8. The armature consists of an anchor ring of iron wire, the strands more or less insulated from each other. Round this anchor ring is wound a continuous endless coil of copper wire; the armature moves in a magnetic field, produced by permanent or electro-magnets with diametrically opposite poles, marked N and S. The lines of magnetic force may be regarded as passing into the ring from N, dividing, passing round the ring and across to S. Thus the coils of wire, both near to N and near to S, are cutting through a very

---

* Of course in applying these conclusions it is necessary to remember that the machines only *tend* to control each other, and that the control of the motive power may be predominant, and *compel* the two or more machines to run at different speeds.

strong magnetic field; consequently there will be an intense inductive action; the inductive action of the coils near *N* being equal and opposite to the inductive action of the coils near *S*, it results that there will be strong positive and negative electric potential at the extremities of a diameter perpendicular to the line *NS*. The electromotive force produced is made use of to produce a current external to the machine thus, the endless coil of the armature is divided into any number of sections, in the diagram into six for convenience, usually into sixty or eighty, and the junction of each pair of sections is connected by a wire to a plate of the commutator fixed upon the shaft which carries the armature; collecting brushes make contact with the commutator as shown in the diagram. If the external resistance were enormously high, so that very little current, or none at all, passed through the armature, the greatest difference of potential between the two brushes would be found when they made contact at points at right angles to the line between the magnets; but when a current passes in the armature, this current causes a disturbing effect upon the magnetic field. Every time the contact of the brushes changes from one contact-plate to the next, the current in a section of the copper coil is reversed, and this reversal has an inductive effect upon all the other coils of the armature. You may take it from me that the net result on any one coil is approximately the same as if that coil alone were moved, and all the other coils were fixed, and there were no reversals of current in them. Now you can easily see that the magnetic effect of the current circulating in the coils of the armature, will be to produce a north pole at *n*, and a south pole at *s*. This will displace the magnetic field in the direction of rotation. If, then, we were to keep the contact points the same as when no current was passing, we should short circuit the sections of the armature at a time when they were cutting through the lines of magnetic force, with

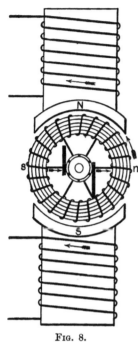

Fig. 8.

a result that there would be vigorous sparks between the collecting brushes and the commutator. To avoid this, the brushes must follow the magnetic field, and also be displaced in the direction of rotation, this displacement being greater as the current in the armature is greater in proportion to the magnetic field. The net effect of this disturbing effect of the current in the armature reacting upon itself is then to displace the neutral points upon the commutator, and consequently somewhat to diminish the effective electromotive force. It is best to adjust the brushes to make contact at a point such that, with the current then passing, flashing is reduced to a minimum, but this point does not necessarily coincide with the point which gives maximum difference of potential. The magnetic field in the Gramme and other continuous dynamo-electric machines, may be produced in several ways. Permanent magnets of steel may be used, as in some of the smaller machines now made, and in all the earlier machines; these are frequently called magneto-machines. Electro-magnets excited by a current from a small dynamo-electric machine, were introduced by Wilde; these may be described shortly as dynamos with separate exciters. The plan of using the whole current from the armature of the machine itself, for exciting the magnets, was proposed almost simultaneously by Siemens, Wheatstone, and S. A. Varley. A dynamo, so excited, is now called a series dynamo. Another method is to divide the current from the armature, sending the greater part into the external circuit, and a smaller portion through the electro-magnet, which is then of very much higher resistance; such an arrangement is called a shunt dynamo. A combination of the two last methods has been recently introduced, for the purpose of maintaining constant potential. The magnet is partly excited by a circuit of high resistance, a shunt to the external circuit, and partly by coils conveying the whole current from the armature. All but the first two arrangements named depend on residual magnetism to initiate the current, and below a certain speed of rotation give no practically useful electromotive force. A dynamo machine is, of course, not a perfect instrument for converting mechanical energy into the energy of electric current. Certain losses inevitably occur. There is, of course, the loss due to friction of bearings, and of the collecting brushes upon the commutator; there is also the loss due to the production of electric currents in the iron of the machine. When these are accounted for, we have the actual electrical effect of the machine in the conducting wire; but all of this is not available for external work. The current has to circulate through the armature, which inevitably has electrical resistance; electrical energy must, therefore, be converted into heat in the armature of the machine. Energy must also be expended in the wire of the electro-magnet which produces the field, for the resistance of this also cannot be reduced beyond a certain limit. The loss by the resistance of the wires of the armature and of the magnets greatly depends on the dimensions of the machine. About this I shall have to say a word or two presently. To know the properties of any machine thoroughly, it is not enough to know its efficiency and the amount of work it is capable of doing; we need to know what it will do under all circumstances of varying resistance or varying electromotive force. We must know, under any given conditions, what will be the electromotive force of the armature. Now this electromotive force depends on the intensity of the magnetic field, and the intensity of the magnetic field depends on the current passing round the electro-magnet and the current in the armature. The current then in the machine is the proper independent variable in terms of which to express the electromotive force. The simplest case is that of the series dynamo, in which the current in the electro-magnet and in the armature is the same, for then we have only one independent variable. The relation between the electromotive force and current is represented by such a curve as is shown in the diagram, Fig. 9 (p. 73). The abscissæ, measured along OX, represent the current, and the ordinates represent the electromotive force in the armature. When four years ago I first used this curve, for the purpose of expressing the results of my experiments on the Siemens dynamo machine, I pointed out that it was capable of solving almost any problem relating to a particular machine, and that it was also capable of giving good indications of the results of changes in the winding of the magnets, or of the armatures of such machines. Since then M. Marcel Deprez has happily named such curves "characteristic curves." I will give you one or two illustrations of their use. A complete characteristic of a series dynamo does not terminate at the origin, but has a negative branch, as shown in the diagram; for it is clear that by reversing the current through the whole machine, the electromotive force is also reversed.

Suppose a series dynamo is used for charging an accumulator, and is driven at a given speed, what current will pass through it? The problem is easily solved. Along OY, Fig. 9, set off OE to represent the electromotive force of the accumulator, and through E draw the line $CEBA$, making an angle with OX,

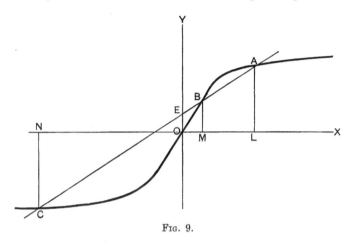

Fig. 9.

such that its tangent is equal to the resistance of the whole circuit, and cutting the characteristic curve, as it in general will do, in three points, A, B, and C. We have then three answers to the question. The current passing through the dynamo will be either OL, OM, or ON the abscissæ of the points where the line cuts the curve. OL represents the current when the dynamo is actually charging the accumulator. OM represents a current which could exist for an instant, but which would be unstable, for the least variation would tend to increase. ON is the current which passes, if the current in the dynamo should get reversed, as it is very apt to do when used for this purpose. The next illustration is rather outside my subject, but shows another method of using the characteristic curve. Many of you have heard of Jacobi's law of maximum effect of transmitting work by dynamo machines. It is this. Supposing that the two dynamo machines were perfect instruments for converting mechanical energy into electrical energy, and that the generating machine were run at constant velocity, whilst the receiving machine had a variable velocity, the greatest amount of work would be developed in the receiving machine when its electromotive force was one-half that of the generating machine, then the efficiency would be one-half, and the electrical work done by the generating machine would be just one-half of what it would be if the receiving machine were forcibly held at rest. Now this law is strictly true if, and only if, the electromotive force of the generating machine is independent of the current passing through its armature. What I am now going to do is to give you a construction for determining the maximum work which can be transmitted when the electromotive force of the generating machine depends on the current passing through the armature, as, indeed, it nearly always does, referring to Fig. 10. OPB is the characteristic curve of the generating machine; construct a derived curve

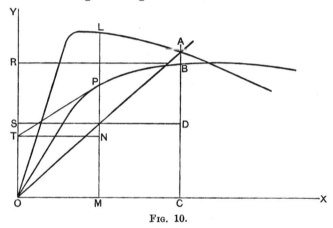

Fig. 10.

thus, at successive points P of the characteristic curve, draw tangents PT, draw TN parallel to OX, cutting PM in N, produce MP to L, making LP equal PN; the point L gives the derived curve, which I want. Now, to find the maximum work which can be transmitted, draw OA at such an angle with OX that its tangent is equal to twice the resistance of the whole circuit, cutting the derived curve in A. Draw the ordinate AC, cutting the characteristic curve in B; bisect AC at D. The work expended upon the generating machine would be represented by the parallelogram $OCBR$, the work wasted in resistance by $OCDS$, and the work developed in the receiving machine by the parallelogram $SDBR$.

When the dynamo-machine is not a series dynamo, but the currents in the armature and in the electro-magnet, though possibly dependent upon each other are not necessarily equal, the problem is not quite so simple. We have, then, two variables, the current in the electro-magnet and the current in the armature; and the proper representation of the properties of the machine will be by a characteristic surface such as that illustrated by this model, Fig. 11. Of the three coordinate axes, OX represents the current in the

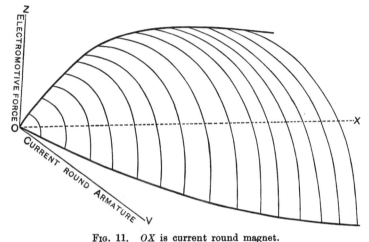

Fig. 11. OX is current round magnet.

magnet; OY represents the current in the armature not necessarily to the same scale, and OZ the electromotive force. By the aid of such a surface as this one may deal with any problem relating to a dynamo-machine, no matter how its electro-magnets and its armature are connected together. Let us apply the model to find the characteristic of a series dynamo. Take a plane through OZ the axis of electromotive force, and making such an angle with the plane OX, OZ, that its tangent is equal to current unity on axis OY, divided by current unity on axis OX. This plane cuts the surface in a curve. The projection of this curve on the plane OX, OZ is the characteristic curve of the series dynamo. This model only shows an eighth part of the complete surface. If any of you should interest yourselves about the other seven parts, which are not without interest, remember that it is assumed that the brushes always make contact with the commutator at the point of no flashing, if there is one. Of course in actual practice one would not use the model of the surface, but the projections of its sections. While I am speaking of characteristic curves there is one point I will just take this opportunity of mentioning. Three years ago Mr Shoolbred exhibited the characteristic curve of a Gramme machine, in which, after the current attained to a certain amount, the electromotive force began to fall. I then said that I thought there must be some mistake in the experiment. However, subsequent experiments have verified the fact; and when one considers it, it is not very difficult to see the explanation. It lies in this: after the current attains to a certain amount the iron in the machines becomes magnetically nearly saturated, and consequently an increase in the current does not produce a corresponding increase in the magnetic field. The reaction, however, between the different sections of the wire on the armature goes on increasing indefinitely, and its effect is to diminish the electromotive force.

A little while ago I said that the dimensions of the machine had a good deal to do with its efficiency. Let us see how the properties of a machine depend upon its dimensions. Suppose two machines alike in every particular excepting that the one has all its linear dimensions double that of the other; obviously enough all the surfaces in the larger would be four times the corresponding surfaces in the smaller, and the weights and volumes of the

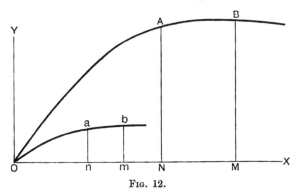

Fig. 12.

larger would be eight times the corresponding weights in the smaller machine. The electrical resistances in the larger machine would be one-half those of the smaller. The current required to produce a given intensity of magnetic field would be twice as great in the larger machine as in the smaller. In the diagram, Fig. 12, are shown the comparative characteristic curves of the two machines, when driven at the same speed. You will observe that the two curves are one the projection of the other, having corresponding points with abscissæ in the ratio of one to two, and the ordinates in the ratio of one to four. Now at first sight it would seem as though, since the wire on the magnet and armature of the larger machine has four times the section of that of the smaller, that four times the current could be carried, that consequently the intensity of the magnetic field would be twice as great, and its area would be four times as great, and hence the electromotive force eight times as great; and since the current in the armature also is supposed to be four times as great, that the work done by the larger machine would be thirty-two times as much as that which would be done by the smaller. Practically, however, no such result can possibly be attained, for a whole series of reasons. First of all, the iron of the magnets becomes saturated, and consequently instead of getting eight times the electromotive force, we should only get four times the electromotive force. Secondly, the current which we can carry in the armature is limited by the rate at which we can get rid of the heat generated in the armature. This we may consider as proportional to its surface, consequently we must only waste four times as much energy in the armature of the larger machine as in the smaller one, instead of eight times, as would be the case if we carried the current in proportion to the section of the wire. Again, the larger machine cannot run at so great an angular velocity as the smaller one. And lastly, since in the larger machine the current in the armature is greater in proportion to the saturated magnetic field than it is in the small one, the displacement of the point of contact of the brushes with the commutator will be greater. However, to cut short the matter, about which one might say a great deal, one may say that the capacity of similar dynamo-machines is pretty much proportionate to their weight, that is, to the cube of their linear dimensions; that the work wasted in producing the magnetic field will be directly as the linear dimensions; and that the work wasted in heating the wires of the armature will be as the square of the linear dimensions. Now let us see how this would practically apply. Suppose we had a small machine capable of producing an electric current of 4 H.-P., that of this 4 H.-P. 1 was wasted in heating the wires of the armature, and 1 in heating the wires of the magnet, 2 would be usefully applied outside. Now, if we doubled the linear dimensions we should have a capacity of 32 H.-P., of which 2 only, if suitably applied, would be required to produce the magnetic field, and 4 would be wasted in heating the wires of the armature, leaving 26 H.-P. available for useful work outside the machine—a very different economy from that of the smaller machines. But if we again doubled the linear dimensions of our machine, we should by no means obtain a similar increase of effect. A consideration of the properties of similar machines has another very important practical use. As you all know, Mr Froude was able to control the design of ironclad ships by experiments upon models made in paraffin wax. Now, it is a very much easier thing to predict what the performance of a large dynamo-machine will be, from laboratory experiments made upon a model of a very small fraction of its dimensions. As a proof of the practical utility of such methods, I may say that by laboratory experiments I have succeeded in increasing the capacity of the Edison machines without increasing their cost, and with a small increase of their percentage of efficiency, remarkably high as that efficiency already was.

I might occupy your time with considerations as to the proper proportion of conductors, and explain Sir W. Thomson's law, that the most economical size of a copper conductor is such that the annual charge for interest and depreciation of the copper of which it is made, shall be equal to the cost of producing the power which is wasted by its resistance. But the remaining time will, perhaps, be best spent in considering the production of light from the energy of electric currents. You all know that this is done commercially in two ways, by the electric arc, and by the incandescent lamp; as the arc lamp preceded the incandescent lamp historically, we will examine one or two points connected with it first.

I have here all that is necessary to illustrate the electric arc, viz., two rods of carbon supported in line with each other, and so mounted that they can be approached or withdrawn. Each carbon is connected with one of the poles of the Edison dynamo-machine, which is supplying electricity to the incandescent lamps which illuminate the whole of this building. A resistance is interposed in the circuit of the lamp, because the electromotive force of the machine is much in excess of what the lamp requires. I now approach the carbons, bring them into contact, and again separate them slightly; you observe that the break does not stop the current which forces its way across the space. I increase the distance between the carbons, and you observe the electric arc between their extremities; at last it breaks, having attained a length of about 1 inch. Now the current has hard work to cross this air-space between the carbons, and the energy there developed is converted into heat, which raises the temperature of the ends of the carbon beyond any other terrestrial temperature. There are several points of interest I wish to notice in the electric arc. Both carbons burn away in the air, but there is also a transference of carbon from the positive to the negative carbon; therefore, although both waste away, the positive carbon wastes about twice as fast as the negative. With a continuous current, such as we are using now, the negative carbon becomes pointed, whilst the positive carbon forms a crater or hollow; it is this crater which becomes most intensely hot and radiates most of the light, hence the light is not by any means uniformly distributed in all directions, but is mainly thrown forward from the crater in the positive carbon. This peculiarity is of great advantage for some purposes, such, for example, as military or naval search-lights, but it necessitates, in describing the illuminating power of an arc-light, some statement of the direction in which the measurement was made. On account of its very high temperature the arc-light sends forth a very large amount of visible radiation, and is therefore very economical of electrical energy. For the same reason its light contains a very large proportion of rays of high refrangibility, blue and ultra-violet. I have measured the red light of an electric arc against the red of a candle, and have found it to be 4,700 times as great, and I have measured the blue of the same arc-light against the blue of the same candle, and found it to be 11,380 times as great. The properties of an electric arc are not those of an ordinary conductor. Ohm's law does not apply. The electromotive force and the current do not by any means bear to each other a constant ratio. Strictly speaking, an electric arc cannot be said to have an electric resistance measurable in ohms. We will now examine the electrical properties of the arc experimentally. In the circuit with the lamp is a Thomson graded current galvanometer for measuring the current passing in ampères; connected to the two carbons is a Thomson graded potential galvanometer, for measuring the difference of potential between them in volts. We have the means of varying the current by varying the resistance, which I have already told you is introduced into the circuit. We will first put in circuit the whole resistance available, and will adjust the carbons so that the distance between them is, so near as I can judge, ⅛ inch. We will afterwards increase the current, and repeat the readings. The results are given in the following Table :—

TABLE III.

Current Galvanometer	Potential Galvanometer	Ampères	Volts	Watts	H.-P.
6·2	12·0	9·9	35	346	0·46
9·3	12·0	14·9	35	521	0·70
11·5	11·8	18·4	34	626	0·84

If the electrical properties of the arc were the same as those of a continuous conductor, the volts would be in proportion to the ampères, if correction were made for change of temperature; you observe that instead of that the potential is nearly the same in the two cases. We may say, with some approach to accuracy, that with a given length of arc, the arc opposes to the current an electromotive force nearly constant, almost independent of the current. This was first pointed out by Edlund. If you will speak of the resistance of the electric arc, you may say that the resistance varies inversely as the current. Take the last experiment; by burning 4 cubic feet of gas per hour we should produce heat-energy at about the same rate. I leave any of you to judge of the comparative illuminating effects. It is not my purpose to describe the mechanisms which have been invented for controlling the feeding of the carbons as they waste away. Several lamps lent by Messrs Siemens Brothers—to whom I am indebted for the lamp and resistance I have just been using—lie upon the table for inspection. An electric arc can also be produced by an alternate current. Its theory may be treated mathematically, and is very interesting, but time will not allow us to go into it. I will merely point out this,—there is some theoretical reason to suppose that an alternate-current arc is in some measure less efficient than one produced by a continuous current. The efficiency of a source of light is greater, as the mean temperature of the radiating surface is greater. The maximum temperature in an arc is limited probably by the temperature of volatilisation of carbon; in an alternate-current arc the current is not constant, therefore the mean temperature is less than the maximum temperature; in a continuous-current arc, the current being constant, the mean and maximum temperatures are equal, therefore in a continuous-current arc the mean temperature is likely to be somewhat higher than in an alternate-current arc.

We will now pass to the simpler incandescent light. When a current of electricity passes through a continuous conductor, it encounters resistance, and heat is generated, as was shown by Joule, at a rate represented by the resistance multiplied by the square of the current. If the current is sufficiently great, the heat will be generated at such a rate that the conductor rises in temperature so far that it becomes incandescent and radiates light. Attempts have been made to use platinum and platinum-iridium as the incandescent conductor, but these bodies are too expensive for general use, and besides, refractory though they are, they are not refractory enough to stand the high temperature required for economical incandescent lighting. Commercial success was not realised until very thin and very uniform threads or filaments of carbon were produced and enclosed in reservoirs of glass, from which the air was exhausted to the utmost possible limit. Such are the lamps made by Mr Edison with which this building is lighted to-night. Let us examine the electrical properties of such a lamp. Here is a lamp intended to carry the same current as those overhead, but of half the resistance, selected because it leaves us a margin of electromotive force wherewith to vary our experiment. Into its circuit I am able to introduce a resistance for checking the current, composed of other incandescent lamps for convenience, but which I shall cover over that they may not distract your attention. As before, we have two galvanometers, one to measure the current passing through the lamp, the other the difference of potential at its terminals. First of all, we will introduce a considerable resistance; you observe that, although the lamp gives some light, it is feeble and red, indicating a low temperature. We take our galvanometer readings. We now diminish the resistance, the lamp is now a little short of its standard intensity; with this current it would last 1000 hours without giving way. We again read the galvanometers. The resistance is diminished still further. You observe a great increase of brightness, and the light is much whiter than before. With this current, the lamp would not last very long. The results are given in the following Table:—

TABLE IV.

Current Galvanometer	Potential Galvanometer	Ampères	Volts	Watts	Resistance Ohms
5·2	12·8	0·38	37	14	97
6·0	14·3	0·44	41	18	93
11·5	23·4	0·84	68	57	81

There are three things I want you to notice in these experiments: first, the light is whiter as the current increases; second, the quantity of light increases very much faster than the power expended increases; and thirdly, the resistance of the carbon filament diminishes as its temperature increases, which is just the opposite of what we should find with a metallic conductor. This resistance is given in ohms in the last column. To the second point, which has been very clearly put by Dr Siemens in his British Association address, I shall return in a minute or two.

The building is this evening lighted by about 200 lamps, each giving sixteen candles' light, when 75 watts of power are developed in the lamp. To produce the same sixteen candles' light in ordinary flat-flame gas-burners, would require between seven and eight cubic feet of gas per hour, contributing heat to the atmosphere at the rate of 3,400,000 foot-pounds per hour, equivalent to 1250 watts, that is to say, equivalent gas lighting would heat the air nearly seventeen times as much as the incandescent lamps.

Look at it another way. Practically, about eight of these lamps take one indicated horse-power in the engine to supply them. If the steam-engine were replaced by a large gas-engine this 1 H.-P. would be supplied by 25 cubic feet of gas per hour, or by rather less; therefore by burning gas in a gas-engine driving a dynamo, and using the electricity in the ordinary way in incandescent lamps, we can obtain more than 5 candles per cubic foot of gas, a result you would be puzzled to obtain in 16-candle gas-burners. With arc lights instead of incandescent lamps many times as much light could be obtained.

At the present time lighting by electricity in London must cost something more than lighting by gas. Let us see what are the prospects of reduction of this cost. Beginning with the engine and boiler, the electrician has no right to look forward to any marked and exceptional advance in their economy. Next comes the dynamo: the best of these are so good, converting 80 per cent. of the work done in driving the machine into electrical work outside the machine, that there is little room for economy in the conversion of mechanical into electrical energy; but the prime cost of the dynamo machine is sure to be greatly reduced. Our hope of greatly increased economy must be mainly based upon probable improvements in the incandescent lamp, and to this the greatest attention ought to be directed. You have seen that a great economy of power can be obtained by working the lamps at high-pressure, but then they soon break down. In ordinary practice, from 140 to 200 candles are obtained from a horse-power developed in the lamps, but for a short time I have seen over 1000 candles per horse-power from incandescent lamps. The problem, then, is so to improve the lamp in details, that it will last a reasonable time when pressed to that degree of efficiency. There is no theoretical bar to such improvements, and it must be remembered that incandescent lamps have only been articles of commerce for about three years, and already much has been done. If such an improvement were realised, it would mean that you would get five times as much light for a sovereign as you can now. As things now stand, so soon as those who supply electricity have reasonable facilities for reaching their customers, electric lighting will succeed commercially where other considerations than cost have weight. We are sure of some considerable improvement in the lamps, and there is a probability that these improvements may go so far as to reduce the cost to one-fifth of what it now is. I leave you to judge whether or not it is probable, nay, almost certain, that lighting by electricity is the lighting of the future.

6.

DYNAMO-ELECTRIC MACHINERY.

[From the *Philosophical Transactions of the Royal Society*, Part I. 1886.]

Received April 19,—*Read May* 6, 1886.

THEORETICAL CONSTRUCTION OF CHARACTERISTIC CURVE.

OMITTING the inductive effects of the current in the armature itself, all the properties of a dynamo machine are most conveniently deduced from a statement of the relation between the magnetic field and the magnetising force required to produce that field, or, which comes to the same thing but is more frequently used in practice, the relation between the electromotive force of the machine at a stated speed and the current around the magnets. This relation given, it is easy to deduce what the result will be in all employments of the machine, whether as a motor or to produce a current through resistance, through an electric arc, or in charging accumulators; also the result of varying the winding of the machine whether in the armature or magnets. The proper independent variable to choose for discussing the effect of a dynamo machine is the current around the magnets, and the primary relation it is necessary to know concerning the machine is the relation of the electromotive force of the armature to the magnet current. This primary relation may be expressed by a curve (Hopkinson, *Mechan. Engin. Instit. Proc.*, 1879, pp. 246 *et seq.*, 1880, p. 266), now called the characteristic of the machine, and all consequences deduced therefrom graphically; or it may be expressed by stating the E.M.F. as an empirical function of the magnetising current. Many such empirical formulæ have been proposed; as an instance we may mention that known as Fröhlich's, according to whom, if c be the current in the magnets, E the resulting E.M.F., $E = \dfrac{ac}{1+bc}$. For some machines this formula is said to express observed results fairly accurately, but in our experience it does not sufficiently approximate to a straight line in the part of the curve near the origin. The character of the error in Fröhlich's formula is apparent by reference to Fig. 1, which gives a series of observations on a dynamo machine, and for comparison therewith a hyperbola F, drawn as favourably as possible to accord with the observations*. Such empirical formulæ possess no advantage over the graphical method aided by algebraic processes, and tend to mask much that is of importance.

One purpose of the present investigation is to give an approximately complete construction of the characteristic curve of a dynamo of given form from the ordinary laws of electro-magnetism and the known properties of iron, and to compare the result of such construction with the actual characteristic of the machine. The laws of electro-magnetism needed are simply (Thomson, papers on *Electrostatics and Magnetism*; Maxwell, *Electricity and Magnetism*, vol. II., pp. 24, 26, and 143), (1) that the line integral of magnetic force around any closed curve, whether in iron, in air, or in both, is equal to $4\pi nc$ where c is the current passing through the closed curve, and n is the number of times it passes through; (2) the solenoidal condition for magnetic induction, that is, if the lines of force or of induction be supposed drawn, then the induction through any tube of induction is the same for every section. Regarding the iron itself, we require to know from experiments on the material in any shape the relation between a, the induction, and α, the magnetic force at any point; for convenience write $a = f^{-1}(\alpha)$, or $\alpha = f(a)$. From these premises, without any further assumption, it is easy to see that a sufficiently powerful and

---

* [Added Aug. 17.—That Fröhlich's formula cannot be a thoroughly satisfactory expression of the characteristic of a dynamo machine is evident from the consideration that E should simply change its sign with c, that is, be an odd function of c. There should be a point of inflexion in the characteristic curve at the origin. Another empirical formula $\dfrac{E}{a} = \tan^{-1}\dfrac{c}{b}$ is free from this objection, but still fails to fully represent the approximation of the curve to a straight line on either side of the origin, and it is equally uninstructive with any other purely empirical formula.]

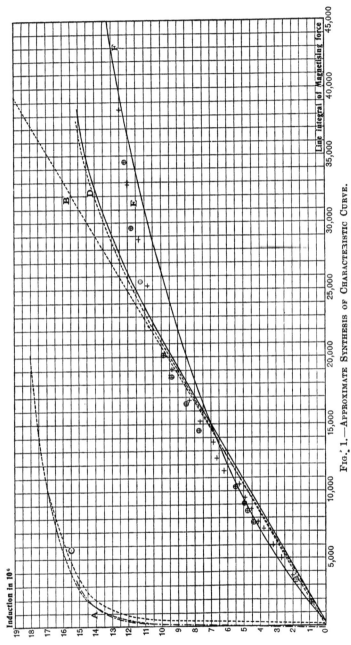

FIG. 1.—APPROXIMATE SYNTHESIS OF CHARACTERISTIC CURVE.

A, armature; B, air space; C, magnets; D, deduced curve; E, observed results, + ascending, ⊕ descending; F, Fröhlich's curve.

laborious analysis would be capable of deducing the characteristic of any dynamo to any desired degree of accuracy. This we do not attempt, as, even if successful, the analysis would not be likely to throw any useful light on the practical problem. We shall calculate the characteristic, first making certain assumptions to simplify matters. We shall next point out the nature of the errors introduced by these assumptions, and make certain small corrections in the method to account for these sources of error, merely proving that the amount of these corrections is probable or deducing it from a separate experiment, and again compare the theoretical and the actual characteristic.

First approximation.—Assume that by some miracle the tubes of magnetic induction are entirely confined to the iron excepting that they pass directly across from the bored faces of the pole-pieces to the cylindrical face of the armature core. This, we shall find, introduces minor sources of error, affecting different parts of the characteristic curve to a material extent. Let I be total induction through the armature, A_1 the area of section of the iron of the armature, l_1 the mean length of lines of force in the armature; A_2 the area of each of the two spaces between core of armature and the pole-pieces of the magnets, l_2 the distance between the core and the pole-piece; A_3 the area of core of magnet, l_3 the total length of the magnets. All the tubes of induction which pass through the armature pass through the space A_2 and the magnet cores, and by our assumption there are no others. We now assume further that these tubes are uniformly distributed over these areas. The induction per square centimetre is then $\dfrac{I}{A_1}$ in the armature core, $\dfrac{I}{A_2}$ in the non-magnetic spaces, $\dfrac{I}{A_3}$ in the magnet cores; the corresponding magnetic forces per centimetre linear must be $f\left(\dfrac{I}{A_1}\right)$, $\dfrac{I}{A_2}$, $f\left(\dfrac{I}{A_3}\right)$. The line integral of magnetic force round a closed curve must be $l_1 f\left(\dfrac{I}{A_1}\right) + 2l_2 \dfrac{I}{A_2} + l_3 f\left(\dfrac{I}{A_3}\right)$. In this approximation we neglect the force required to magnetise pole-pieces and other parts not within the magnet coils to avoid complication. The equation of the characteristic curve is then $4\pi n c = l_1 f\left(\dfrac{I}{A_1}\right) + 2l_2 \dfrac{I}{A_2} + l_3 f\left(\dfrac{I}{A_3}\right)$.

This curve is, of course, readily constructed graphically from the magnetic property of the material expressed by the curve $\alpha = f(a)$. In Fig. 1 curve A represents $x = l_1 f\left(\dfrac{y}{A_1}\right)$, the straight line B $x = 2l_2 \dfrac{y}{A_2}$, curve C $x = l_3 f\left(\dfrac{y}{A_3}\right)$, and curve D the calculated characteristic. When we compare this with an actual characteristic E, we shall see that, broadly speaking, it deviates from truth in two respects: (1) it does not rise sufficiently rapidly at first; (2) it attains a higher maximum than is actually realised. Let us examine these errors in detail.

(1) The angle the characteristic makes with the axis of abscissæ near the origin is mainly determined by the line B. We have in fact a very considerable extension of the area of the field beyond that which lies under the bored face of the pole-piece. The following consideration will show that the extension may be considerable. Imagine an infinite plane slab, and parallel with it

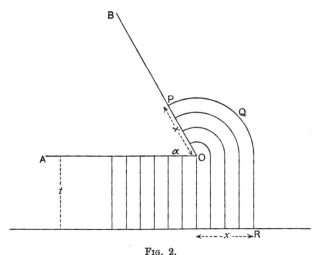

Fig. 2.

a second slab cut off by a second plane making an angle α. We want a rough idea of the extension of the area between the plates by the spreading of the lines of induction beyond the boundary. We know that the actual extension of the area will be greater than we shall calculate it to be if we prescribe an arbitrary distribution of lines of force other than that which is consistent with Laplace's equation.

Assume, then, the lines of force to be segments of circles centre O, and straight lines perpendicular to OA. The induction along a line PQR will be $\dfrac{V}{(\pi-\alpha)x+t}$, V being difference of potential between the planes, and the added induction from OPB will be $\displaystyle\int_0^x \dfrac{V\,dx}{(\pi-\alpha)x+t} = \dfrac{V}{\pi-\alpha}\log\dfrac{(\pi-\alpha)x+t}{t}$. Thus if $\alpha = \dfrac{\pi}{2}$ we have for $x = t, 2t$, &c.

x	$\dfrac{1}{\pi-\alpha}\log\dfrac{(\pi-\alpha)x+t}{t}$
t	·599
$2t$	·904
$3t$	1·109
$4t$	1·263
$5t$	1·387
$10t$	1·792

showing that the extension of the area of the field is likely to be considerable.

(2) The failure of the actual curve to reach the maximum indicated by approximate theory is because the theory assumes that all tubes of induction passing through the magnets pass also through the armature. Familiar observations round the pole-pieces of the magnets show that this is not the case. If ν be the ratio of the total induction through the magnets to the induction in the armature we must, in our expression for the line integral of magnetising force, replace the term $f\left(\dfrac{I}{A_3}\right)$ by $f\left(\dfrac{\nu I}{A_3}\right)$; ν is not strictly a constant, as we shall see later; it is somewhat increased as I increases owing to magnetisation of the core of the armature, and it is also affected by the current in the armature. For our present purpose we treat it as constant.

There is yet another source of error which it is necessary to examine. Some part of the induction in the armature may pass through the shaft instead of through the iron plates. An idea of the amount of this disturbance may be readily obtained. Consider the closed curve $ABCDEF$, AB and $FEDC$ are drawn along lines of force, AF and BC are orthogonal to lines of force. Since this closed curve has no currents passing through it, the line integral of force around it is nil; therefore, neglecting force along ED, we have force along AB is equal to force along FE and DC. In the machine presently described we may safely neglect the induction through the shaft; the error is comparable with the uncertainty as to the value of l_1; but in another machine, with magnets of much greater section, the effect of the shaft would become very sensible when the core is practically saturated.

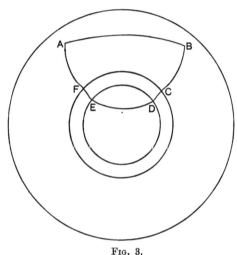

Fig. 3.

The amended formula now becomes

$$4\pi nc = l_1 f\left(\frac{I}{A_1}\right) + 2l_2\frac{I}{A_2} + l_3 f\left(\frac{\nu I}{A_3}\right) + l_4 f\left(\frac{\nu I}{A_4}\right) + 2l_5 f\left(\frac{I}{A_5}\right)$$

where l_4 is the mean length of lines of force in the wrought-iron yoke, A_4 the area of the yoke, l_5 and A_5 corresponding quantities for the pole-pieces, the last two terms being introduced for the forces required to magnetise the yoke and the two pole-pieces.

We now repeat the graphical method of construction exactly as before, the actual observations of induction in armature and current being plotted on the same diagram (Fig. 1), in which curve G represents the force required to magnetise the yoke, and curve H that required to magnetise the pole-pieces. Before discussing these curves further, and comparing the results with those of actual observation, it may be convenient to describe the machine upon which the experiments have been made, confining the description strictly to so much as is pertinent to our present inquiry.

DESCRIPTION OF MACHINE.

The dynamo has a single magnetic circuit, consisting of two vertical limbs, extended at their lower extremities to form the pole-pieces, and having their upper extremities connected by a yoke of rectangular section. Each limb, together with its pole-piece, is formed of a single forging of wrought-iron. These forgings, as also that for the yoke, are built up of hammered scrap and afterwards carefully annealed, and have a magnetic permeability but little inferior to the best Swedish charcoal iron. The yoke is held to the limbs by two bolts, the surfaces of contact being truly planed. In section the limb is oblong, with the corners rounded in order to facilitate the winding of the magnetising coils. A zinc base, bolted to the bed-plate of the machine, supports the pole-pieces.

The magnetising coils are wound directly on the limbs, and consist of 11 layers on each limb of copper wire 2·413 mms. diameter (No. 13, B.W.G.), making 3260 convolutions in all, the total length being approximately 4570 metres. The pole-pieces are bored to receive the armature, leaving a gap above and below, subtending an angle of 51° at the centre of the fields. The opposing surfaces of the gap are 8 mms. deep.

The following table gives the leading dimensions of the machine :—

	cms.
Length of magnet limb	= 45·7
Width of magnet limb	= 22·1
Breadth of magnet limb	= 44·45
Length of yoke	= 61·6
Width of yoke	= 48·3
Depth of yoke	= 23·2
Distance between centres of limbs	= 38·1
Bore of fields	= 27·5
Depth of pole-piece	= 25·4
Width of pole-piece measured parallel to the shaft	= 48·3
Thickness of zinc base	= 12·7
Width of gap	= 12·7

The armature is built up of about 1000 iron plates, insulated one from another by sheets of paper, and held between two end plates, one of which is secured by a washer shrunk on to the

92 DYNAMO-ELECTRIC MACHINERY.

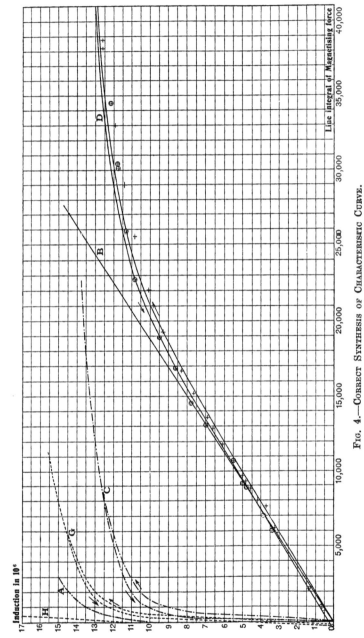

FIG. 4.—CORRECT SYNTHESIS OF CHARACTERISTIC CURVE.

A, armature; *B*, air space; *C*, magnets; *D*, calculated curve; *E*, observations, + ascending, ⊕ descending; *G*, yoke; *H*, pole-piece.

shaft, and the other by a nut and lock-nut screwed on the shaft itself. The plates are cut from sheets of soft iron, having probably about the same magnetic permeability as the magnet cores. The shaft is of Bessemer steel, and is insulated before the plates are threaded on.

The following table gives the leading dimensions of the armature:—

	cms.
Diameter of core	= 24·5
Diameter of internal hole	= 7·62
Length of core over the end plates	= 50·8
Diameter of shaft	= 6·985

The core is wound longitudinally according to the Hefner von Alteneck principle with 40 convolutions, each consisting of 16 strands of wire 1·753 mm. diameter, the convolutions being placed in two layers of 20 each. The commutator is formed of 40 copper bars, insulated with mica, and the connections to the armature so made that the plane of commutation in the commutator is horizontal, when no current is passing through the armature.

Fig. 5 shows a side elevation of the dynamo; Fig. 6 a cross-section through the centres of the magnets; Fig. 7 a section of

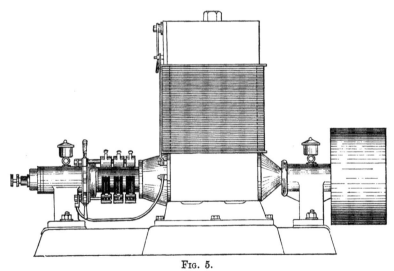

FIG. 5.

the core of the armature, in a plane through the axis of the shaft.

94 DYNAMO-ELECTRIC MACHINERY.

The dynamo is intended for a normal output of 105 volts 320 ampères at a speed of 750 revolutions per minute. The resistance

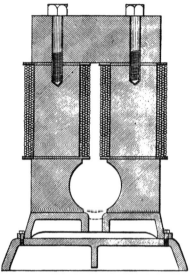

FIG. 6.

of the armature measured between opposite bars of the commutator is 0·009947 ohm, and of the magnet coils 16·93 ohms, both at a temperature of 13·5° Centigrade; Lord Rayleigh's determination of the ohm being assumed.

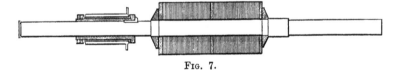

FIG. 7.

We have now to estimate the lengths and areas required in the synthesis of the characteristic curve.

A_1;—from the length of the core of the armature (50·8 cms.) must be deducted 3·4 cms. for the thickness of insulating material between the plates; the resultant area is, on the other hand, as has already been stated, slightly augmented by the presence of the steel shaft. A_1 is taken as 810 sq. cms.

l_1;—this is assumed to be 13 cms.: i.e., slightly in excess of the shortest distance (12·6 cms.) between the pole-pieces.

A_2;—the angle subtended by the bored face of the pole-piece at the axis is 129°, the breadth of the pole-piece is 48·3 cms., the diameter of the bore of the field is 27·5 cms., and, as already stated, the diameter of core 25·5 cms., thus, the area of pole-piece is 1513 sq. cms., and the area of 129° of the cylinder at the mean radius of 13·0 cms. is 1410 sq. cms.; this value is taken for A_2 in the curves drawn in Fig. 1. In Fig. 4, A_2 is taken as 1600, an allowance of 190 sq. cms. being made for the spreading of the field at the edges of the pole-pieces, or $\frac{190}{160} = 1·2$ cm. all round the periphery, that is $\frac{1·2}{1·5} = 0·8$ of the distance from iron of pole-pieces to iron of core.

l_2 is 1·5 cm.

A_3 is a little uncertain, as the forgings are not tooled all over, it is here taken as 980 sq. cms., but this value may be slightly too high.

l_3 is 91·4 cms.

A_4 is 1120 sq. cms.

l_4 is 49 cms., being measured along a quadrant from the centre of the magnet, thus:—

Fig. 8.

A_5 is 1230 sq. cms., intermediate between the area of magnet and face of pole-piece.

l_5 is 11 cms.

ν was determined by experiment as described below, and its value is taken as 1·32; when the magnetising current is more than 5·62 ampères its value should be a little greater.

The function $f(a)$ is taken from Hopkinson, *Phil. Trans.*, vol. CLXXVI. (1885), p. 455*; the wrought-iron there referred to was not procured at the same time as, and its properties may differ to a certain extent from, the wrought-iron of these magnets.

The curves now explain themselves; the abscissæ in each case represent the line integral of magnetising force in the part of

---

* The material parts of this paper are printed at the end of this volume for convenience of reference.

the magnetic circuit referred to, the ordinates, the number of lines of induction which also pass through the armature.

The results of the actual observations on the machine are indicated, those when the magnetising force is increasing +, when it is decreasing ⊕. The measurements of the currents in the magnets which were separately excited, and of the potential difference between the brushes, the circuit being open, were made with Sir W. Thomson's graded galvanometers, standardised at the time of use. The irregularities of the observations are probably due to the variation of speed, the engine being not quite perfectly governed. The second construction exhibits quite as close an agreement between observation and calculation as could be expected; the deviation at high magnetising forces is probably due to three causes, increase in the value of ν when the core of the armature is partially saturated, uncertainty as to the area A_3, difference in the quality of the iron. It is interesting to see how clearly theory predicts the difference between the ascending and descending curves of a dynamo. Consideration of the diagram proves that this machine is nearly perfect in its magnetic proportions. The core might be diminished by increasing the hole through it to a small, but very small, extent without detriment. Any reduction of area of magnets would be injurious; they might, indeed, be slightly increased with advantage. An increase in the length of the magnets would be very distinctly detrimental. Again, little advantage results from increasing the magnetising force beyond the point at which the permeability of the iron of the magnets begins to rapidly diminish. For iron of the same quality as that of the machine under consideration, a magnetising force of $2·6 \times 10^3$ or $28·4$ per centimetre is suitable. To get the same induction in other parts of the circuit, the diagram shows that for the air space a magnetising force of 21×10^3 is required, for the pole-pieces $0·1 \times 10^3$, for the armature $0·2 \times 10^3$, for the yoke $0·6 \times 10^3$, making a total force required of $24·5 \times 10^3$. Any alteration in the length or the area of any portion of the magnetic circuit entails a corresponding alteration in the magnetising forces required for that portion, at once deducible from the diagram. Similar machines must have the magnetising forces proportional to the linear dimensions, and, consequently, if the electromotive force of the machines is the same, the diameter of the wire of the magnet coils must be proportional to the linear dimensions. If the lengths of the several portions of the magnetic circuit remain the same, but the areas are similarly altered, the section of the wire must be altered in proportion to the alteration in the periphery of the section.

EXPERIMENT TO DETERMINE ν.

Around the middle of one of the magnet limbs a single coil of wire was taken, forming one complete convolution, and its ends connected to a Thomson's mirror galvanometer rendered fairly ballistic. If the circuit of the field magnets, while the exciting current is passing, be suddenly short-circuited, the elongation of the galvanometer is a measure of the total induction within the core of the limbs, neglecting the residual magnetisation. If the short circuit be suddenly removed, so that the current again passes round the field-magnets, the elongation of the galvanometer will be equal in magnitude and opposite in direction.

The readings taken were:

Zero . . . . . 71 left.
Deflection . . . . 332 „ magnets made.
„ . . . . 196 right; magnets short-circuited.
Hence, deflection to right = 267
„ left = 261
Mean deflection = 264

To determine the induction through the armature, the leads to the ballistic galvanometer were soldered to consecutive bars of the commutator, connected to that convolution of the armature, which lay in the plane of commutation.

The readings taken were:

Zero . . . . . . 23 left
Deflection . . . . . 223 „ magnets made.
„ . . . . . 176⎫ right; magnets
„ . . . . . 178⎭ short-circuited.
Hence, deflection to right and left = 200

It thus appears that out of 264 lines of force passing through the cores of the magnet limbs at their centre, 200 go through the core of the armature, whence ν equals 1·32. The magnetising current round the fields during these experiments was 5·6 ampères.

EXPERIMENTS ON WASTE FIELD NOT PASSING THROUGH ARMATURE.

As in the determination of ν a single convolution was taken around the middle of one of the limbs, and connected to the ballistic galvanometer; the deflections, when a current of 5·6 ampères was suddenly passed through the fields or short-circuited, were:

Zero . . . . . 34 left.
Deflection . . . 148 „ magnets made.
„ . . . . 82 right; magnets short-circuited.
Hence, deflection to right = 116
„ left = 114
Mean deflection = 115

I. Four convolutions were then wound round the zinc plate and the cast-iron bed in a vertical plane, passing through the axis of the armature; and the deflections noted. The mean deflection was 50·25; or, reducing to one convolution, = 12·6.

II. A square wooden frame, 38 cms. × 38 cms., on which were wound ten convolutions, was then inserted between the magnet limbs, with one side resting on the armature, and an adjacent side projecting 5 cms. beyond the coils on the limbs, or about 7·6 cms. beyond the cores of the limbs. The mean deflection was 55; or, reducing to one convolution, = 5·5.

III. The same frame was raised a height of 6·35 cms. above the armature in a vertical plane. The mean deflection was 46·5 or, reducing to one convolution, = 4·6.

IV. The same frame was again lowered on the armature and pushed inwards so as to lie symmetrically within the space between the limbs. The deflections were noted and the mean was 80; or, reducing to one convolution, = 8·0.

Let G represent the leakage through a vertical area bounded by the armature, and a line 7·6 cms. above the armature, and of the same width as the pole-pieces; let R be the remainder of the leakage between the limbs; then II. and III. give

$$\tfrac{2}{3}G + \tfrac{2}{3}R = 5\cdot5$$
$$\tfrac{2}{3}R = 4\cdot6$$

whence
$$G = 1{\cdot}35$$
$$R = 6{\cdot}9.$$

Again, IV. gives
$$\tfrac{5}{6}(G + R) = 8{\cdot}0$$
therefore
$$G + R = 9{\cdot}6$$

which shows an agreement as near as might be expected considering the rough nature of the experiment, and that the leakage is assumed uniform over the areas considered.

We take
$$G = 1{\cdot}6$$
$$R = 8{\cdot}0.$$

Reducing these losses to percentages we have

$G = \dfrac{1{\cdot}6}{115}$ . . . . = 1·4 per cent.

$R = \dfrac{8{\cdot}0}{115}$ . . . . = 7·0 ,,

And from I. the leakage through the zinc plate and iron base . . . . } = 10·3 ,,

Hence the two gaps account for . . 2·8 ,,
The zinc plate and iron base account for . 10·3 ,,
And the area between the limbs . . 7·0 ,,
Making a total loss accounted for . . 20·1 ,,
Out of an observed loss of . . . 24·24 ,,

The leakage through the shaft and from pole-piece to yoke, and one pole-piece to the other by exterior lines, will account for the remainder.

Effect of the Current in the Armature.

The currents in the fixed coils around the magnets are not the only magnetising forces applied in a dynamo machine; the currents in the moving coils of the armature have also their effect on the resultant field. There are in general two independent variables in a dynamo machine, the current around the magnets and the current in the armature, and the relation of E.M.F. to currents is fully represented by a surface. In well-constructed machines the effect of the latter is reduced to a minimum, but it can be by no means neglected. When a section of the armature coils is commutated it must inevitably be momentarily short-circuited, and if at the time of commutation the field in which the section is moving is other than feeble, a considerable current will arise in that section, accompanied by waste of power and destructive sparking. It may be well at once to give an idea of the possible magnitude of such effects. In the machine already described the mean E.M.F. in a section of the armature at a certain speed may be taken as 6 volts, its resistance 0·000995 ohm. Setting aside, then, for the moment questions of self-induction, if a section were commutated at a time when it was in a field of one-tenth part of the mean intensity of the whole field there would arise in that section, whilst short-circuited by the collecting brush, a current of 600 ampères, four times the current when the section is doing its normal work. The ideal adjustment of the collecting brushes is such that during the time they short-circuit the sections of the armature the magnetic forces shall just suffice to stop the current in the section, and to reverse it to the same current in the opposite direction.

Suppose the commutation occurs at an angle λ in advance of the symmetrical position between the fields, and that the total current through the armature be C, reckoned positive in the direction of the resultant E.M.F. of the machine, i.e., positive when the machine is used as a generator of electricity. Taking any closed line through magnets and armature, symmetrically drawn as $ABCDEFA$ (fig. 9), it is obvious that the line integral of magnetic force is diminished by the current in the armature included between angle λ in front and angle λ behind the plane of symmetry. If m be the number of convolutions of the armature, the value of this magnetising force is $4\pi C \dfrac{m}{2} \dfrac{2\lambda}{\pi} = 4\lambda m C$ opposed to the magnetising force of the fixed coils on the magnets. Thus, if we know the lead of the brushes and the current in the armature we are at once in a position to calculate the effect on the electromotive force of the machine. A further effect of the current in the armature is a material disturbance in the distribution of the induction over the bored face of the pole-piece; the force along BC (fig. 9) is by no means equal to that along

DE. Draw the closed curve *BCGHB*, the line integral along *CG* and *HB* is negligible. Hence the difference between force *HG* and *BC* is equal to $4\pi C \dfrac{m}{2} \dfrac{\kappa}{\pi} = 2\kappa m C$ where κ is the angle *COG*. This disturbance has no material effect upon the performance of

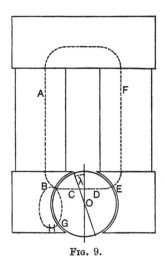

Fig. 9.

the machine. But the current in the armature also distorts the arrangement of the comparatively weak field in the gap between the pole-pieces, displacing the point of zero field in the direction of rotation in a generator and opposite to the direction of rotation in a motor; and it is due to this that the non-sparking point for the brushes is displaced. A satisfactory mathematical analysis of the displacement of the field in the gap between the pole-pieces by the current in the armature would be more troublesome than an à priori analysis of the distribution of field in this space when the magnet current is the only magnetising force. Owing to the fact that the armature is divided into a finite number of sections there is a rapid diminution of the displacement of the field during the time that a section is being commutated, the diminution being recovered whilst the brush is in contact with only one bar of the commutator. The field thus oscillates slightly, owing to the disturbance caused by reversing the direction of the current in the successive sections of the armature. The number of oscillations in a gramme armature or in a Siemens' armature with an even number of sections will be pm, where p is the number of revolutions per second, but in a Siemens' armature with an odd number of sections it will be $2pm$.* This oscillation of the field is only another way of expressing the effect of the self-induction of the section, but it must be remembered that if the self-induction, multiplied by change of current, is expressed as a change in the field we must omit self-induction as a separate term in our electrical equations. The precise lead to be given to the brushes in order to avoid sparking in any given case depends on many circumstances—the form and extent of the pole-pieces, the number of sections in the armature, and the duration of the short circuit which the brushes cause in any section of the armature. The adjustment of the position of the collecting brushes is generally made by hand at the discretion of the attendant, and is in some cases fixed once for all to suit an average condition of the machine. We shall, therefore, treat λ the lead as an independent variable, controlled by the attendant.

Let I be total induction through the armature, $I + I'$ total induction through the magnets, I' being the waste field. Let C be current in armature, c in the magnets. Let gI' be the line integral of magnetic force from a point on one pole-piece to a point on the other, the line being drawn external to the armature, g will be approximately constant. Omitting as comparatively unimportant the magnetising force in the pole-pieces and iron core of the armature we have the following equations:—

$$4\lambda mC + 2l_2 \frac{I}{A_2} - gI' = 0$$

$$4\lambda mC + 2l_2 \frac{I}{A_2} + l_3 f\left(\frac{I + I'}{A_3}\right) = 4\pi nc.$$

---

* [Added Aug. 17.—Armatures with an odd number of convolutions are open to one theoretical objection, which would be a practical one if the number of convolutions were very small. The $2m+1$ convolutions constitute in themselves a closed circuit, having a resistance four times the mean actual resistance of the armature measured between the collecting brushes. When any one convolution is exactly in the middle of the field, the E.M.F. of the other $2m$ convolutions exactly balance, so that there is upon the closed circuit an E.M.F. due to the single convolution somewhat in excess of $\frac{1}{m}$th part of the actual E.M.F. of the machine. Thus there will be an alternating E.M.F. around the closed circuit of the armature capable of causing a considerable waste of power. This waste is materially checked by the self-induction of the circuit.]

When $C = 0$ we observed

$$I = \frac{1}{\nu - 1} I',$$

whence

$$g = \frac{1}{\nu - 1} \frac{2l_2}{A_2},$$

eliminating I'

$$\frac{2l_2}{\nu A_2} \left(\nu I + 4\lambda m C \frac{A_2}{2l_2}(\nu - 1) \right) + l_3 f \left\{ \frac{\nu I + 4\lambda m C \cdot \frac{A_2}{2l_2}(\nu - 1)}{A_3} \right\}$$

$$= 4\pi nc + 4\lambda m C \frac{\nu - 1}{\nu} - 4\lambda m C = 4\pi nc - 4\lambda m C \frac{1}{\nu}.$$

The characteristic curve when $C = 0$ being $I = F(4\pi nc)$ we may write the above as the equation of the characteristic surface thus

$$I + \frac{\nu - 1}{\nu} 4\lambda m C \frac{A_2}{2l_2} = F\left(4\pi nc - \frac{4\lambda m C}{\nu} \right).$$

In applying this equation it must not be forgotten that the E.M.F. of the machine cannot be determined from I unless the

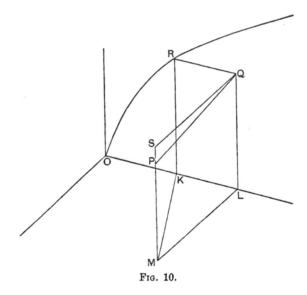

Fig. 10.

commutation occurs at such a time that the coil being commutated embraces all, or nearly all, the lines of induction in the armature.

This equation enables the characteristic surface to be constructed from the characteristic curve. Let $OL = 4\pi nc$, $LM = 4m\lambda C$, draw MK so that $\dfrac{KL}{LM} = \dfrac{1}{\nu}$, through K draw ordinate KR meeting characteristic curve in R, draw RQ parallel to OL meeting ordinate QL in Q, draw QS parallel to LM; draw QP so that $\dfrac{PS}{SQ} = \dfrac{\nu-1}{\nu} \cdot \dfrac{A_2}{2l_2}$. Then P is a point on the characteristic surface.

A very important problem is to deduce the characteristic curve of a series wound machine from the normal characteristic; in this case $c = C$, and we have

$$I + \frac{\nu-1}{\nu} 4\lambda mC \frac{A_2}{2l_2} = F\left\{\left(4\pi n - \frac{4\lambda m}{\nu}\right) C\right\}$$

taking PR as ordinate of any point in the normal characteristic, cut off QR equal to $\dfrac{\nu-1}{\nu} 4\lambda mC \dfrac{A_2}{2l_2}$, that is, draw OQ so that

$$\tan QOx = \frac{\nu-1}{\nu} 4\lambda mC \frac{A_2}{2l_2} \Big/ 4\pi \left(n - \frac{m\lambda}{\nu\pi}\right) C$$

$$= \frac{\nu-1}{\nu} \frac{A_2}{2l_2} \frac{\lambda m}{\pi n - \dfrac{\lambda m}{\nu}},$$

then PQ will represent the induction corresponding to magnetising force $4\pi \left(n - \dfrac{m\lambda}{\nu\pi}\right) C$. It is noteworthy that as the current C, and therefore OR increases, PQ, the induction, will attain a maximum and afterwards diminish, vanish, and become negative. That in series wound machines the E.M.F. has a maximum value has been many times observed. The cause lies in the existence of a waste field not passing through the armature, and in the saturation of the magnet core.

The effect of the current in the armature on the potential between the brushes of any machine is the same as that of an addition to the resistance of the armature proportional to the lead of the brushes, and to the ratio of the waste field to the total field, combined with that of taking the main current $\dfrac{m\lambda}{\nu\pi}$ times round the magnets in direction opposite to the current c. The preceding investigation tells the whole story of a dynamo machine, excepting only the relation of λ to C in order that the brushes may be so placed as to avoid sparking. The only constant or function which has to be determined experimentally for any particular machine is ν, the ratio of total to effective field, all the rest follows from the configuration of the iron and the known properties of the material.

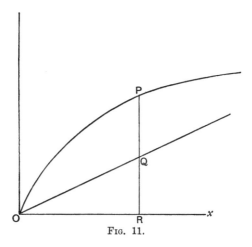

FIG. 11.

The following illustrations of the effect of the current in the armature and the lead of the brushes are interesting. In both cases the magnet coils are supposed to be entirely disconnected so that c is zero. First, let λ be negative, short-circuit the brushes and drive the machine at a certain speed, a large current will be produced, the current in the armature itself forming the magnet[*]. Second, let λ be positive, cause a current to pass through the armature, the armature will turn in the positive direction and will

* [Added Aug. 17.—This experiment was tried upon a dynamo machine of construction generally similar to that shown in Figs. 5, 6, and 7, but with an armature of half the length intended in normal work to give 400 ampères, 50 volts, at 1,000 revolutions. The magnet coils were disconnected, and the terminals of the armature were connected through a Siemens' electrodynamometer, and the machine was run at 1,380 revolutions. When the brushes were placed in the normal position ($\lambda = 0$) the current due to residual magnetism was 52 ampères. By giving the brushes a small positive lead the current was reduced to nearly zero. By giving the brushes a small negative lead a current of over 234 ampères, the maximum measured by the dynamometer, was obtained, and by varying the lead it was easy to maintain a steady current of any desired amount.]

act as a motor capable of doing work. In either case, particularly the former, such use of the machine would not be practical owing to violent sparking on the commutator. The following is a further illustration of the formula given above. If we could put up with the sparking which would ensue, it would be possible to make λ negative in a generator of electricity, and thereby obtain by the reactions of the armature itself all the results usually obtained by compound winding.

Efficiency Experiments.

Having discussed the relations subsisting between the configuration of the magnetic circuit of a dynamo machine and the induction obtained for given magnetising forces, and having compared the results obtained by direct calculation with the results of actual observation on a particular machine, the construction of which we have described at length, it appeared of importance to determine the efficiency of the machine under consideration as a converter of energy, when used either as a generator of electricity or as a motor. An accurate determination of the mechanical power transmitted to a dynamo by a driving belt, or of the power given by a motor presents formidable experimental difficulties. Moreover, if the mechanical power absorbed in driving the dynamo be measured directly, any error in measurement will involve an error of the same magnitude in the determination of the efficiency. To avoid this difficulty we employed the following device.

Let two dynamos, approximately equal in dimensions and power, have their shafts coupled by a suitable coupling, which may serve also as a driving pulley; and let the electrical connexions between the dynamos be made so that the one drives the other as a motor. If the combination be driven by a belt passing over the coupling pulley, the power transmitted by the belt is the waste in the two dynamos and the connexions between them. By suitably varying the magnetic field of one of the dynamos, the power passing between the two machines can be adjusted as desired. If, then, the electrical power given out by the generator is measured, and also the power transmitted by the belt, the efficiency of the combination can be at once determined. By this arrangement the measurement, which presents experimental difficulties, viz., of the power transmitted by the belt, is of a small quantity. Consequently, even a considerable error in the determination has but a small effect on the ultimate result. On the other hand, the measurement of the large quantity involved, viz., the electrical power passing between the two machines, can without difficulty be made with great accuracy.

The second machine was similar in all respects to that already described, and each is intended for a normal output of 110 volts, 320 ampères, at a speed of 780 revolutions per minute.

The power transmitted by the belt was measured by a dynamometer of the Hefner-Alteneck type, the general arrangement being as shown in the diagram, Fig. 12. A is the driving pulley of the engine, B the driven coupling of the dynamos; D, D are

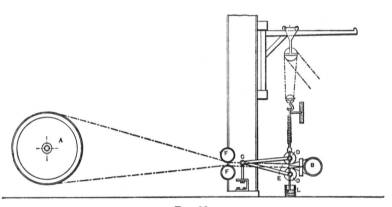

Fig. 12.

the guide pulleys of the dynamometer carried on a double frame turning about the fulcrum C, and supported by a spiral spring, the suspension of which can be varied by a pair of differential pulley-blocks attached to a fixed support overhead. When a reading is made the suspension of the spring is adjusted until the index of the dynamometer comes to a fiducial mark on a fixed scale, the extension of the spring is then read by a second index attached to its upper extremity; F, F are two fixed guide pulleys of the same diameter as the pulleys D, D, and having the same distance between their centres, in order that the two portions of the belt may be parallel and the sag as far as possible taken up. The normal from C to the centre line of either portion of the belt between the pulley A and the guide pulleys $= 31\cdot 9$ cms. The normal from C to the centre line of either of the parallel portions of the belt $= 2\cdot 4$ cms.; and from C to the centre line of the spring $= 92\cdot 7$ cms.

Take moments about C; then

$$\text{Tension of the belt} = \frac{92\cdot 7}{34\cdot 3} \times \text{tension of spring},$$

$$= 2\cdot 7 \times \text{tension of spring}.$$

Also the diameter of the pulley $B = 33\cdot 6$ cms. and the thickness of the belt $= 1\cdot 6$ cm.

Hence the velocity of the centre of the belt in centimetres per second $= 1\cdot 845 \times$ revolutions of dynamo per minute, and, therefore,

Power transmitted by the belt in ergs per second

$= 2\cdot 7 \times 1\cdot 845 \times 981 \times$ tension of spring $\times$ revolutions per minute,

assuming the value of g to be 981.

We may more conveniently express the power in watts ($= 10^7$ ergs per second), and write

Power in watts

$= \cdot 0004887 \times$ tension of spring $\times$ revolutions per minute.

The potential between the terminals of the generator was measured by one of Sir William Thomson's graded galvanometers, previously standardised by a Clark's cell, which had been compared with other Clark's cells, of which the electromotive force was known by comparison with Lord Rayleigh's standard. The current between the two machines was measured by passing it through a known resistance, the difference of potential between the ends of the resistance being determined by direct comparison with the Clark's standard cell, according to Poggendorff's method. As experiments were made with currents of large magnitude, it was important that the temperature coefficient of the resistance should be as low as possible. To this end we found a resistance-frame constructed of platinoid wire of great value. The temperature coefficient of this alloy is only $0\cdot 021$ per cent. per degree Centigrade. (*Proc. Roy. Soc.*, vol. 38, p. 265 (1885).)

The resistances of the armatures and magnets of the two machines are as follows:—

			Ohms.
Generator	. .	armature . .	0·009947
		magnets . .	16·93
Motor . . .		armature . .	0·009947
		magnets . .	16·44.

The resistance of the leads connecting the two machines was 0·00205 ohm, and of the standard resistance 0·00586 ohm.

In all determinations of resistance, the value of the B.A. ohm was taken as $0·9867 \times 10^9$ c.g.s. units, according to Lord Rayleigh's determination.

The diagram shows the electrical connexions between the two machines with the rheostat r inserted in the magnets of the motor dynamo.

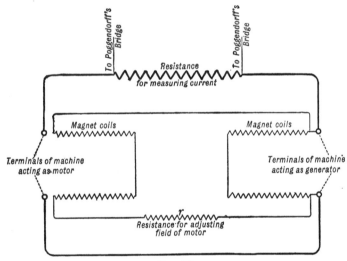

Fig. 13.

In order to ascertain the friction of bending the belt round the pulley B, and of the journals of the dynamo, a preliminary experiment was made with the dynamometer. The combination was run at a speed of 814 revolutions per minute with the dynamos on open circuit and the tension of the spring observed— 9979 grammes. The engine was then reversed, and the dynamos run at the same speed and the tension of the spring again

110   DYNAMO-ELECTRIC MACHINERY.

observed—3629 grammes. The difference of the two readings gives twice the power absorbed in friction, viz.: 1262 watts for the two machines, or 631 watts per machine. This is excluded entirely from the subsequent determinations of efficiency, as being a quantity dependent on such arbitrary conditions as the lubrication of the journals, the weight of the belt, and the angle it makes with the horizontal.

In Table I. column I. is the speed of the dynamos; column II. is the reading of the spring in grams; column III. is the power transmitted by the belt in watts; column IV. is the potential at the terminals of the generator; column V. is the current passing

TABLE I.

	I. Revolutions per minute	II. Grams.	III. Watts	IV. Volts	V. Ampères	VI. Ohms	VII. Watts
1	810	8,392	3,322	129·1	21·6	1·39	13
2	801	9,299	3,640	127·2	72·0	1·39	75
3	811	11,113	4,405	125·8	150·0	2·72	267
4	808	10,433	4,119	124·4	186·0	2·72	397
5	792	10,660	4,124	116·5	211·0	2·72	499
6	798	16,897	6,589	110·6	351·0	4·59	1,309
7	764	17,690	6,605	110·12	358·0	4·09	1,360
8	766	17,804	6,665	110·6	360·0	4·59	1,375
9	778	16,556	6,294	102·3	369·0	4·09	1,436
10	756	20,412	7,541	96·8	446·0	4·59	2,070
11	808	9,526	3,765	119·3	36·8	2·72	25
12	802	3,855	1,512	113·5	No current	......	......
13	814	3,175	1,262	......	......	......	......

	VIII. Watts	IX. Watts	X. Watts	XI. Watts	XII. Watts	XIII. Watts	XIV. Watts
1	4	984	861	77	4,720	691	5,411
2	51	955	837	112	11,096	805	11,901
3	223	935	709	295	20,896	988	21,884
4	344	914	695	388	25,256	691	25,949
5	443	801	608	453	26,590	660	27,250
6	1,222	722	455	1,101	41,433	890	42,323
7	1,268	716	473	1,131	42,087	828	42,915
8	1,289	722	455	1,152	42,494	836	43,330
9	1,354	618	408	1,178	40,314	650	40,964
10	1,979	554	348	1,670	46,244	459	46,704
11	13	841	637	116	5,998	1,066	7,064
12	......	......	......	......	......	756	......
13	......	......	......	......	......	631	......

in the external circuit between the two machines; column VI. is the resistance introduced into the magnets of the motor by the rheostat; column VII. is the power absorbed in the armature of the generator; column VIII. is the power absorbed in the armature of the motor; column IX. is the power absorbed in the magnets of the generator; column X. is the power absorbed in the magnets of the motor; column XI. is the power absorbed in the connecting leads between the two dynamos, in the rheostat resistance r, and in the standard resistance used for measuring the current; column XII. is the total electrical power developed in the generator; column XIII. is half the power absorbed by the combination less the known losses in the armatures, magnets, and external connexions of the two machines; column XIV. is the total mechanical power given to the generator, being the sum of the powers given in columns XII. and XIII.

In Table II. the percentage losses in the armature and magnets of the generator are given, as also the sum of all other losses as obtained from column XIII. in the foregoing table; also the percentage efficiency of the generator, of the motor, and of the

TABLE II.

	I. Per cent.	II. Per cent.	III. Per cent.	IV. Per cent.	V. Per cent.	VI. Per cent.
1	0·24	18·20	12·76	68·8	57·28	39·40
2	0·63	8·93	6·76	84·58	82·99	70·19
3	1·22	4·27	4·52	90·00	90·15	81·13
4	1·53	3·52	2·66	92·28	92·65	85·49
5	1·83	2·94	2·415	92·80	93·12	86·42
6	3·09	1·71	2·10	93·10	93·30	86·86
7	3·17	1·67	1·93	93·23	93·39	87·07
8	3·17	1·67	1·93	93·23	93·43	87·10
9	3·51	1·75	1·59	93·39	93·50	87·32
10	4·43	1·19	0·98	93·39	93·36	87·19
11	0·35	11·9	15·1	72·65	65·77	47·78

double conversion. Column I. is the percentage loss in the generator armature; column II. is the percentage loss in the generator magnets; column III. is the percentage sum of all other losses in the generator; column IV. is the percentage efficiency of the generator; column V. is the percentage efficiency of the motor; column VI. is the percentage efficiency of the double conversion.

In this series of experiments in all cases, from Nos. 1 to 10 inclusive, the brushes, both of the generator and motor, were set at the non-sparking point; but in No. 11 no lead was given to the brushes of the generator, and, consequently, there was violent sparking throughout the duration of the experiment.

In No. 12 the magnets were separately excited with a current giving 113·5 volts across their terminals. The power absorbed must be due entirely to local currents in the core of the armature, and the energy for the reversal of magnetisation of the core twice in every revolution of the armature.

No. 13 gives the results of the experiments on the friction of the bearings and in bending the belt already referred to.

It will be observed that the figures in column XIII. are calculated by deducting the power absorbed in the armatures and magnets, and extraneous resistances from the total power given to the combination as measured by the dynamometer. They must therefore include all the energy dissipated in the core of the armature, whether in local currents or in the reversal of its magnetisation; also the energy dissipated in local currents in the pole-pieces, if such exist; also the energy spent in reversing the direction of the current in each convolution of the armature as they are successively short circuited by the brushes. Further, it will include the waste in all the connexions of the machine from the commutator to its terminals and the friction of the brushes against the commutator. A separate experiment was made to determine the amount of this last constituent, but it was found to be too small to be capable of direct measurement by the dynamometer. Moreover, from the manner in which the figures in this column are deduced, any error in the dynamometric measurement will appear wholly in them. Since, undoubtedly, the first two components enumerated are the most important, and the conditions determining their amount are practically the same throughout the series, the close agreement of the figures in the column are a fair criterion of the accuracy of the observations. Probably 100 watts is the limit of error in any of the measurements. Such an error would affect the determination of the efficiency when the machines were working up to their full power to less than $\frac{1}{4}$ per cent.

It has been assumed that the sum of these losses is equally

divided between the two machines. This will not accurately represent the facts, as the intensities of the fields and the currents passing through the armatures differ to some extent in the two machines. The inequality, however, cannot amount to a great quantity, and if it diminishes the efficiency of the generator it will increase the efficiency of the motor by a like amount, and contrariwise. In No. 11 of the series the effect of the sparking at the brushes of the generator is very marked, the power wasted amounting to at least 250 watts.

If it be assumed that the dissipation of energy is the same whether the magnetisation of the core is reversed by diminishing and increasing the intensity of magnetisation without altering its direction, or whether it is reversed by turning round its direction without reducing its amount to zero, a direct approximation may be made to the value of this component. (J. Hopkinson, *Phil. Trans.*, vol. 176 (1885), p. 455.)

The core has about 16,400 cubic centims. of soft iron plates, hence loss in magnetising and demagnetising when the speed is 800 revolutions per minute $= 16,400 \times \frac{800}{60} \times 13,356$ ergs per second $= 292$ watts.

Referring to Table II. it appears that the efficiency approaches a maximum when the current, passing externally between the two machines, is about 400 ampères. Let C be the current in the armature, ρ its resistance, W the power absorbed in all parts of the machine other than the armature, then, if the speed is constant, the efficiency is approximately $\frac{EC - W - C^2\rho}{EC}$, where E is the electromotive force. This is a maximum when $\frac{W}{C} + C\rho$ is a minimum, which occurs when $W = C^2\rho$; when the loss in the armature is equal to the sum of all other losses. For the machines under consideration the experimental results verify this deduction. But in actual practice the rate of generation of heat in the armature conductors, when a current of 400 ampères was passed for a long period, would be so great as to trench upon the margin of safety desirable in such machines. Of the total space, however, available for the disposition of the conductors only about onefourth part is actually occupied by copper, the remainder being taken up with insulation, and the interstices left by the round wire. If the space occupied by the copper could be increased to three-fourths of the total space available, while the cooling surface remained the same, the current could be increased 75 per cent. and the efficiency increased 1·3 per cent. approximately, as all losses other than that in the armature-wires would not be materially altered.

The loss in the magnets is also susceptible of reduction. It has already been shown that for a given configuration of the magnetic circuit and a given electromotive force the section of the wire of the magnet coils is determinate. The length is, however, arbitrary, since within limits the number of ampère convolutions is independent of the length. An increase in the length will cause a proportionate diminution in the power absorbed in the magnet coils. If the surface of the magnets is sufficient to dissipate all the heat generated, then the length of wire is properly determined by Sir William Thomson's rule that the cost of the energy absorbed must be equal to the continuing cost of the conductor.

APPENDIX.

(Added Aug. 17.)

Since the reading of the present communication experiments have been tried on machines having armatures wound on the plan of Gramme and with differently arranged magnets; the experiments were carried out in a closely similar manner to that already described.

DESCRIPTION OF MACHINES.

The construction of these machines is shown in figs. 14, 15, and 16, in which fig. 14 shows an elevation, fig. 15 a section through the magnets, fig. 16 a longitudinal section of the armature. It will be observed that the magnetic circuit is divided. The pole-pieces are of cast-iron and are placed above and below the armature, and are extended laterally. The magnet cores are of wrought-iron of circular section and fit into the extensions of the cast-iron pole-pieces, so that the area of contact of the cast-iron is greater than the area of section of the magnet. The magnetising coils consist of 2196 convolutions on each limb of copper wire, No. 17, B.W.G., in No. 1 machine, and 2232 convolutions in No. 2

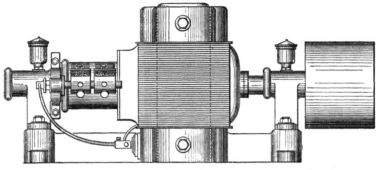

Fig. 14.

machine. The pole-pieces are bored to receive the armature, leaving a gap on either side subtending an angle of 41° at the axis.

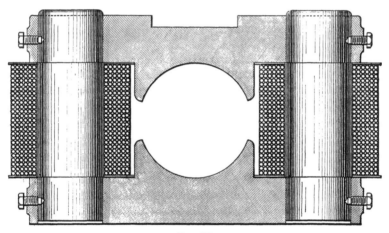

Fig. 15.

The bearings are carried upon an extension of the lower pole-piece.

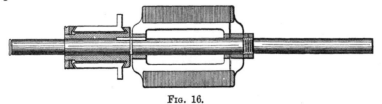

Fig. 16.

116   DYNAMO-ELECTRIC MACHINERY.

The following table gives the principal dimensions of the magnets in No. 1 machine:—

	cms.
Length of magnet limbs between pole-pieces	26·0
Diameter of magnet limb	15·24
Bore of fields	25·7
Width of pole-piece parallel to the shaft	24·1
Width of gap between poles	8·6

The armature is built up of plates as in the machine already described, and is carried from the shaft by a brass frame between the arms of which the wires pass.

The principal dimensions are as follows:—

	cms.
Diameter of core	24·1
Diameter of hole through core	14·0
Length of core over end plates	24·1

The core is wound on Gramme's principle with 160 convolutions, each consisting of a single wire, No. 10, B.W.G., the wire lying on the outside of the armature in a single layer. The commutator has 40 bars.

This dynamo is compound wound, and is intended for a normal output of 105 volts 130 ampères, at a speed of 1050 revolutions per minute. The resistance of the armature is 0·047 ohm, and of the magnet shunt coils 53·7 ohms.

There is here no yoke, and consequently A_4 and l_4 do not appear in the equation.

It is necessary to bear in mind that the magnetising force is that due to the convolutions on one limb, and that the areas are the sums of the areas of the two limbs. In calculating induction from E.M.F. it is also necessary to remember that two convolutions in a Gramme count as one in a Hefner-Alteneck armature.

A_1;—the section of the core is 245 sq. cms., allowances for insulation reduce this to 220·5 sq. cms.

l_1;—this is assumed to be 10 cms., but it will be seen that an error in this value has a much more marked effect on the characteristic in this machine than in the other.

A_2;—the angle subtended by the bored face of the pole-pieces is 139°, the mean of the radii of the pole-pieces and the core is 12·45 cms. Hence the area of 139° of the cylinder of this radius is 768·3 sq. cms., add to this a fringe of a width 0·8 of the distance from core to pole-pieces as already found necessary for the other machine, and we have 839·5 sq. cms. as the value of A_2.

l_2 is 0·8 cm.

A_3 is 365 sq. cms. (i.e., the area of two magnet cores).

l_3 is 26·0 cms.

A_5 is taken to be 955 sq. cms., viz., double the smallest section of the pole-piece.

l_5 is a very uncertain quantity; it is assumed to be 15 cms.

The expression already used requires slight modification. Inasmuch as the pole-pieces are of cast-iron a different function must be used. Different constants for waste field must be used for the field, the pole-pieces and the magnet core. We write

$$4\pi nc = l_1 f\left(\frac{I}{A_1}\right) + 2l_2 \frac{\nu_2 I}{A_2} + l_3 f\left(\frac{\nu_3 I}{A_3}\right) + 2l_5 f'\left(\frac{\nu_5 I}{A_5}\right)$$

the function f' is taken from Hopkinson, *Phil. Trans.*, vol. CLXXVI. (1885), p. 455, Plate 52. ν_2, ν_3, and ν_5 were determined by experiment as described below, their values are

$$\nu_2 = 1·05$$
$$\nu_3 = 1·18$$
$$\nu_5 = 1·49.$$

Comparing the curves in Fig. 4 with that in Fig. 17, the most notable difference is that in the present case the armature core is more intensely magnetised than the magnet cores. No published experiments exist giving the magnetising force required to produce the induction here observed in the armature core, amounting to a maximum of 20,000 per sq. cm. We might, however, make use of such experiments as the present to construct roughly the curve of magnetisation of the material; thus we find that with this particular sample of iron a force of 740 per cm. is required to produce induction 20,000 per sq. cm.; this conclusion must be regarded as liable to considerable uncertainty.

The observations on the two machines are plotted together, but are distinguished from each other as indicated. They are unfortunately less accurate than those of fig. 4, and are given here merely as illustrating the method of synthesis.

118 DYNAMO-ELECTRIC MACHINERY.

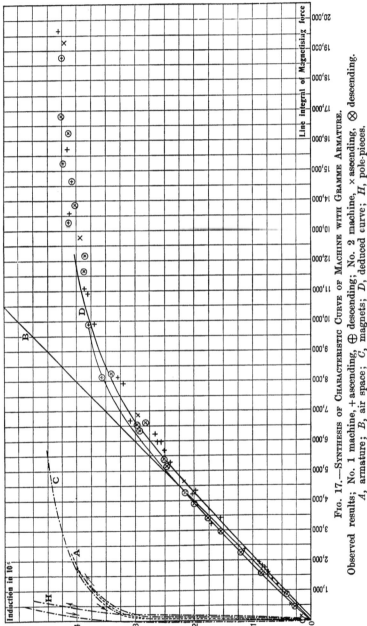

FIG. 17.—SYNTHESIS OF CHARACTERISTIC CURVE OF MACHINE WITH GRAMME ARMATURE.
Observed results; No. 1 machine, + ascending, ⊕ descending; No. 2 machine, × ascending, ⊗ descending. A, armature; B, air space; C, magnets; D, deduced curve; H, pole-pieces.

Experiments to Determine ν_2, ν_3, and ν_5.

The method was essentially the same as is described on pp. 97 and 98, and was only applied to No. 1 machine. Referring to fig. 18 a wire AA was taken four times round the middle of one limb of the magnet, a known current was suddenly passed round the magnets, the elongation of the reflecting galvanometer

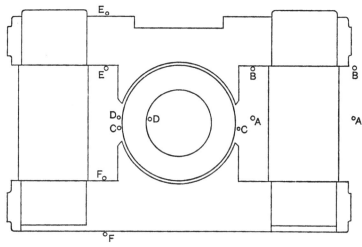

Fig. 18.

was observed, it was found to be 214 scale divisions, giving 107 as the induction through the two magnet limbs in terms of an arbitrary unit. The coil was moved to the top of the limb as at BB—the elongation was reduced to 206 or 103 for the two limbs. We take the mean induction in the magnet to be 105. A wire was taken three times round the whole armature in a horizontal plane as at CC; the elongation observed was 222 divisions or 74 in terms of the same units. A wire was taken four times round one-half of the armature as at DD; the elongation was 141, or induction in the iron of the armature 70·5, whence we have

$$\nu_2 = \frac{74}{70\cdot5} = 1\cdot05.$$

It may be well to recall here that ν_2 is essentially dependent on the intensity of the field, strictly the line B in fig. 17 should not be straight but slightly curved.

Four coils were taken round the upper pole-piece at *EE*; the elongation was 159, giving 79·5 on the two sides. Coils at *FF* give a higher result, 87·5, owing to the lines of induction which pass round by the bearings of the machine, and across to the upper ends of the magnets—ν_5 is taken to be $\dfrac{83\cdot5}{70\cdot5} = 1\cdot18$.

Efficiency Experiments.

The method and instruments were those already described, pp. 106 and 107, excepting that the current was measured by a Thomson's graded galvanometer, which had been standardised against a Clark's cell in the position and at the time when used. The resistance of leading wires and galvanometer was 0·034 ohm, the series coils introduced for compounding the machines were also brought into use, and the losses due to their resistance find a place in columns XII. and XIII. of the following Table III.,

Table III.

	I. Degrees	II. Revolutions	III. Grams	IV. Watts	V. Volts	VI. Ampères	VII. Ohms	VIII. Watts
1	17·5	1098	7711	4419	100·1	139·0	∞	955
2	5	1094	2722	1554	103·8	41·2	18·8	105
3*	0	1144	1814	1083	104·7	7·85	0	11

	IX. Watts	X. Watts	XI. Watts	XII. Watts	XIII. Watts	XIV. Watts	XV. Watts	XVI. Watts	XVII. Watts
1	895	372	0	497	464	657	16,395	289	16,684
2	78	400	138	55	41	148	5,015	294	5,309
3*	3	395	409	6	1	2	1,637	128	1,765

in which column I. is lead of brushes of the dynamo, positive for the generator, negative for the motor; column II., revolutions per minute; column III., deflection of spring in grammes; column IV., watts by dynamometer; column V., volts at terminals of generator;

---

* In this experiment the direction of the current had become reversed, and No. 2 machine was generator.

column VI., ampères in external circuit; column VII., rheostat resistance; column VIII., watts in generator armature; column IX., watts in motor armature; column X., watts in generator shunt magnet coils; column XI., watts in motor shunt; column XII., watts in generator series magnet coils; column XIII., watts in motor series; column XIV., watts in external resistances; column XV., total electrical power of generator; column XVI., half the sum of losses unaccounted for; column XVII., total mechanical power applied to generator.

TABLE IV.

	Generator Armature	Generator Shunt Coils	Generator Series Coils	Other Losses	Efficiency of Generator	Efficiency of Motor	Efficiency of Double Conversion
1	5·8	2·2	3·0	1·9	87·1	89·0	77·5
2	2·0	7·5	1·0	5·5	84·0	92·0	77·3

Table IV. gives the losses and efficiencies as percentages in exactly the same way as in Table II., excepting that another column is introduced for the loss in the series coils of the magnets of the generator.

[The core of the armature contains about 6500 cub. cms. of iron. Hence energy of magnetising and demagnetising when the speed $= 1100$ revolutions per minute $= 6500 \times \frac{1100}{60} \times 13,356$ in ergs per second $= 159$ watts.]

In conclusion we desire to express our indebtedness to Messrs. Mather and Platt, by whom the machines we have used were manufactured, and who, in placing the same at our disposal, together with all facilities for carrying out our experiments at their Salford Iron Works, have enabled us to put theory to the test of experiment on an engineering scale.

7.

DYNAMO-ELECTRIC MACHINERY*.

[From the *Proceedings of the Royal Society*, Vol. LI.]

Received February 15, 1892.

THE following is intended as the completion of a Paper† by Drs J. and E. Hopkinson (*Phil. Trans.*, 1886)†. The motive is to verify by experiment theoretical results concerning the effect of the currents in the armature of dynamo machines on the amount and distribution of the magnetic field which were given in that Paper, but which were left without verification. For the sake of completeness, part of the work is given over again.

The two dynamos experimented upon were constructed by Messrs Siemens Brothers and Co., and are identical as far as it is possible to make them. They are mounted upon a common base plate, their axles being coupled together, and are referred to in this Paper respectively as No. 1 and No. 2.

Each dynamo has a single magnetic circuit consisting of two vertical limbs extended at their lower extremities to form the pole-pieces, and having their upper extremities connected by a yoke of rectangular section. Each limb, together with its pole-piece, is formed of a single forging of wrought-iron. These forgings, as also that of the yoke, are built up of hammered scrap iron, and afterwards carefully annealed. Gun-metal castings bolted to the base plate of the machine support the magnets.

* It must not be supposed from his name not appearing in this short Paper that my brother, Dr E. Hopkinson, had a minor part in the earlier Paper. He not only did the most laborious part of the experimental work, but contributed his proper share to whatever there may be of merit in the theoretical part of the Paper.— J. H.

† The Paper here referred to is that reprinted on pages 84 to 121 of this volume.

The magnetizing coils on each limb consist of sixteen layers of copper wire 2 mms. in diameter, making a total of 3,968 convolutions for each machine. The pole-pieces are bored out to receive the armature, leaving a gap above and below subtending an angle of 68° at the centre of the shaft. The opposing surfaces of the gap are 1·4 cm. deep.

The following table gives the leading dimensions of the machine:—

	cms.
Length of magnet limb	66·04
Width of magnet limb	11·43
Breadth of magnet limb	38·10
Length of yoke	38·10
Width of yoke	12·06
Depth of yoke	11·43
Distance between centres of limbs	23·50
Bore of fields	21·21
Depth of pole-piece	20·32
Thickness of gun-metal base	10·80
Width of gap	12·06

The armature core is built up of soft iron disks, No. 24 B. W. G., which are held between two end plates screwed on the shaft.

The following table gives the leading dimensions of the armature:—

	cms.
Diameter of core	18·41
Diameter of shaft	4·76
Length of core	38·10

The core is wound longitudinally according to the Hefner von Alteneck principle with 208 bars made of copper strip, each 9 mms. deep by 1·8 mm. thick. The commutator is formed of fifty-two hard drawn copper segments insulated with mica, and the connections to the armature so made that the plane of commutation in the commutator is vertical when no current is passing through the armature.

Each dynamo is intended for a normal output of 80 ampères, 140 volts, at 880 revolutions per minute. The resistance of the armature measured between opposite bars of the commutator is 0·042 ohm, and of each magnet coil 13·3 ohms.

In the machine the armature core has a greater cross section than the magnet cores, and consequently the magnetizing force used therein may be neglected. The yoke has the same section as the magnet cores, and is therefore included therein, as is also the pole-piece. The formula connecting the line integral of the magnetizing force and the induction takes the short form

$$4\pi nc = 2l_2 \frac{I}{A_2} + l_3 f\left(\frac{\nu I}{A_3}\right)*,$$

where

n is the number of turns round magnet;

c is the current round magnet in absolute measure;

l_2 the distance from iron of armature to rim of magnet;

A_2 the corrected area of field;

I the total induction through armature;

l_3 the mean length of lines of magnetic force in magnets;

A_3 the area of section of magnets;

ν the ratio of induction in magnets to induction in armature;

f the function which the magnetizing force is of the induction in the case of the machine actually taken from Dr J. Hopkinson on the "Magnetization of Iron," *Phil. Trans.*, 1885, Figs. 4 and 5, Plate 47.

In estimating A_2 we take the mean of the diameter of the core and of the bore of the magnets 19·8 cms., and the angle subtended by the pole face 112°, and we add a fringe all round the area of the pole face equal in width to the distance of the core from the pole face. This is a wider fringe than was used in the earlier experiments†, because the form of the magnets differs slightly. The area so estimated is 906 sq. cms.

l_3 is taken to be 108·8 cms.

A_3 is 435·5 sq. cms.

ν was determined by the ballistic galvanometer to be 1·47. It is to be expected that, as the core is actually greater in area than

---

* *Phil. Trans.* 1886; page 90 of this volume.
† *Phil. Trans.* 1886; page 95 of this volume.

the magnets, ν will be more nearly constant than in the earlier experiments. It was found to be constant within the limits of errors of observation.

Referring to Fig. 1, the curve C is the curve $x = l_3 f\left(\dfrac{\nu y}{A_3}\right)$, and the straight line B is the curve $x = 2l_2 \dfrac{y}{A_2}$, while the full line D is the characteristic curve of the machine,

$$x = 2l_2 \frac{y}{A_2} + l_3 f\left(\frac{\nu y}{A_3}\right),$$

as given by calculation.

The marks + indicate the results of actual observations on machine No. 1, and the marks 0 the results on machine No. 2, the total induction I being given by the equation:—

$$I = \frac{\text{potential difference in volts} \times 10^8}{208 \times \text{revolutions per second}}.$$

Experiments made upon the power taken to drive the machine under different conditions show that it takes about 250 watts

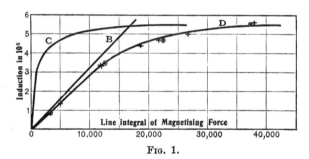

Fig. 1.

more power to turn the armature at 660 revolutions when the magnets are normally excited than when they are not excited at all. The volume of the core is 9,465 cub. cms., or in each complete cycle the loss per cubic centimetre is

$$\frac{250 \times 10^7}{11 \times 9,465} = 24,000 \text{ ergs.}$$

The loss by hysteresis is about 13,000 (*Phil. Trans.*, 1885, p. 463) if the reversals are made by variation of intensity of the magnetizing force and the iron is good wrought-iron. This result is similar to that in the earlier Paper*, where it is shown that the actual loss in the core, when magnetized, is greater than can be accounted for by the known value of hysteresis.

EFFECTS OF THE CURRENT IN THE ARMATURE.

Quoting from the Royal Society Paper [page 99 of this volume], "The currents in the fixed coils around the magnets are not the only magnetizing forces applied in a dynamo machine—the currents in the moving coils of the armature have also their effect on the resultant field. There are in general two independent variables in a dynamo machine—the current around the magnets and the current in the armature; and the relation of E.M.F. to currents is fully represented by a surface. In well-constructed machines the effect of the latter is reduced to a minimum, but it can be by no means neglected. When a section of the armature coils is commutated, it must inevitably be momentarily short circuited; and if at the time of commutation the field in which the section is moving is other than feeble, a considerable current will arise in that section, accompanied by waste of power and destructive sparking.......

"Suppose the commutation occurs at an angle λ in advance of the symmetrical position between the fields, and that the total current through the armature be C, reckoned positive in the direction of the resultant E.M.F. of the machine, *i.e.*, positive when the machine is used as a generator of electricity. Taking any closed line through magnets and armature, symmetrically drawn as *ABCDEFA* [Fig. 9, p. 101], it is obvious that the line integral of magnetic force is diminished by the current in the armature included between angle λ in front and angle λ behind the plane of symmetry. If m be the number of convolutions of the armature, the value of this magnetizing force is $4\pi C \dfrac{m}{2} \dfrac{2\lambda}{\pi} = 4\lambda m C$ opposed to the magnetizing force of the fixed coils on the magnets. Thus if we know the lead of the brushes and the current in the armature we are at once in a position to calculate the effect on the electromotive force of the machine. A further effect of the current in the armature is a material disturbance of the distribution of the induction over the bored face of the pole-piece; the force along *BC* [Fig. 9] is by no means equal to that along *DE*.

* See page 113 of this volume.

Draw the closed curve $BCGHB$, the line integral along CG, and HB is negligible. Hence the difference between force HG and BC is equal to $4\pi C \dfrac{m}{2} \dfrac{\kappa}{\pi} = 2\kappa mC$, where κ is the angle COG."

To verify this formula is one of the principal objects of this Paper.

A pair of brushes having relatively fixed positions near together, and insulated from the frame and from one another, are carried upon a divided circle, and bear upon the commutator. The difference of potential between these brushes was measured in various positions round the commutator, the current in the armature, the potential difference of the main brushes, and the speed of the machine being also noted.

The results are given in Figs. 2, 3, 4, and 5, in which the ordinates are measured potential differences, and the abscissæ are angles turned through by the exploring brushes. The potential differences in Fig. 2 were measured by a Siemens voltmeter, and each ordinate is therefore somewhat smaller than the true value, owing to the time during which the exploring brushes were not actually in contact with the commutator segments. But this does not affect the results, because the area is reduced in the same proportion as the potential differences. In Figs. 3, 4, and 5 the potential differences were taken on one of Sir William Thomson's quadrant electrometers, and are correct.

Take Fig. 2, in which machine No. 1 is a generator. A centimetre horizontally represents 10° of lead*, and the ordinates

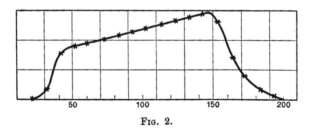

FIG. 2.

represent differences of potential between the brushes. The area of the curve is 61·3 sq. cms., and represents 130 volts and a total

* In this paper the diagrams are on half the scale of the original, so that 10° of lead is really represented by half a centimetre horizontally.

field of $\dfrac{130}{104} \times \dfrac{1}{29} \times 10^8 = 4\cdot31 \times 10^6$ lines of induction. This is, of course, not the actual field, which is 3 per cent. greater on account of the resistance of the armature, but is represented by an area 3 per cent. greater. An ordinate of 1 cm. will represent an induction of $\dfrac{4\cdot31}{61\cdot3} \times 10^6 = 7\cdot0 \times 10^4$ lines in 10°. The area of 10° is $39\cdot5 \times 1\cdot73 = 68\cdot3$ sq. cms.* Hence an ordinate of 1 cm. represents an induction of 1,024 lines per square centimetre. The difference between ordinates at 50° and 140° is 2·5; hence the difference of induction is actually 2,560. Theoretically, we have $\kappa = \tfrac{1}{2}\pi$, $m = 104$, $C = 9\cdot4$. Therefore $2\kappa mC = 3{,}072$, and this is the line integral of magnetizing force round the curve.

Let A be the induction at 50° and $A + \delta$ at 140°: these also are the magnetizing forces. Hence $(A + \delta)\,1\cdot4 - A\,1\cdot4 = 2\kappa mC$; $\delta = 2{,}200$, as against 2,560 actually observed.

Take Fig. 3, in which No. 2 machine is a motor. The total field $= \dfrac{107}{104} \times \dfrac{1}{20} \times 10^8 = 5\cdot15 \times 10^6$ lines of induction. Since

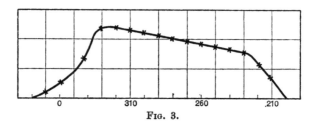

Fig. 3.

the area of the diagram is 53·5 sq. cms., an ordinate of 1 cm. $= \dfrac{5\cdot15}{53\cdot5} \times 10^6 = 9\cdot6 \times 10^4$ lines of induction in 10°. Hence an ordinate of 1 cm. represents an induction of $\dfrac{9\cdot6 \times 10^4}{68\cdot3} = 1{,}400$ lines per square centimetre. The difference between ordinates at 320° and at 230° is 2·0; hence the difference of induction is actually 2,800. Theoretically, we have $\dfrac{2\kappa mC}{l} = \dfrac{3\tfrac{1}{4} \times 104 \times 11\cdot4}{1\cdot4} = 2{,}666$, as against 2,800 actually observed.

* In calculating this area, the allowance for fringe at ends of armature is taken less than before, because the form of opposing faces differs.

In Fig. 4, No. 1 machine is a generator. The total field
$= \frac{52}{104} \times \frac{1}{12\cdot 6} \times 10^8 = 3\cdot 97 \times 10^6$ lines. The area of the diagram
is 90·9 sq. cms., and therefore an ordinate of 1 cm. $= \frac{3\cdot 97}{90\cdot 9} \times 10^6$
$= 4\cdot 37 \times 10^4$ lines in 10°. Hence an ordinate of 1 cm. represents

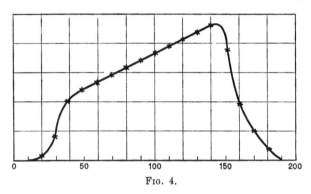

FIG. 4.

an induction of $\frac{4\cdot 37 \times 10^4}{68\cdot 3} = 639$ lines per square centimetre.
The difference between ordinates at 50° and at 140° is 4·5; hence
the difference of induction is actually 2,877. Theoretically, we
have $\frac{2\kappa mC}{l} = \frac{3\frac{1}{7} \times 104 \times 12\cdot 9}{1\cdot 4} = 3,010$, as against 2,877.

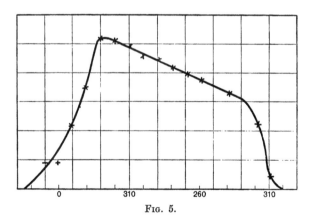

FIG. 5.

In Fig. 5, No. 2 machine is a motor. The total field
$= \frac{63\cdot 5}{104} \times \frac{1}{12\cdot 3} \times 10^8 = 4\cdot 96 \times 10^6$ lines. The area of the diagram is

112·2 sq. cms., and therefore an ordinate of 1 cm. $= \dfrac{4\cdot 96}{112\cdot 2} \times 10^6$ $= 4\cdot 42 \times 10^4$ lines in 10°. Hence an ordinate of 1 cm. represents an induction of $\dfrac{4\cdot 42}{68\cdot 3} \times 10^4 = 647$ lines per square centimetre. The difference between ordinates at 323° and at 233° is 4·2; hence the difference of induction is actually 2,718. Theoretically, we have $\dfrac{2\kappa mC}{l} = \dfrac{3\frac{1}{7} \times 104 \times 12\cdot 3}{1\cdot 4} = 2{,}870$, as against 2,718 actually observed.

At page 103 of the preceding Paper on Dynamo-Electric Machinery it is shown that

$$I + \dfrac{\nu - 1}{\nu} 4\lambda mC \dfrac{A_2}{2l_2} = F\left(4\pi nc - \dfrac{4\lambda mC}{\nu}\right),$$

where $I = F(4\pi nc)$ is the characteristic curve when $C = 0$, and λ is the lead of the brushes.

The following is an endeavour to verify this formula. The potentials both upon the magnets and upon the brushes were taken by a Siemens voltmeter, and are rough. The speeds were taken by a Buss tachometer, and there is some uncertainty about the precise lead of the brushes, owing to the difficulty in determining the precise position of the symmetrical position between the fields, and also to the width of the contacts on the commutator.

It was necessary, in order to obtain a marked effect of the armature reaction, that the magnet field should be comparatively small, that the current in the armature should be large, and the leads of the brushes should be large.

The two machines had their axles coupled so that No. 1 could be run as a generator, and No. 2 as a motor. The magnets were in each case coupled parallel, and excited by a battery each through an adjustable resistance. The two armatures were coupled in series with another battery, and the following observations were made:—

	Potential on Magnets in volts	Potential on Brushes	Speed per Minute	Current in Ampères	Lead of Brushes
No. 1	24—24	66—67	880	102—103	26°
No. 2	29—29	86—84	880	102—103	29°

From which we infer:—

	Current in Magnets	$4\pi nc$	Corrected Potential for Resistance of Armature	Total Induction, I
No. 1	1·78	8,900	70·8	$2·30 \times 10^6$
No. 2	2·15	10,750	80·7	$2·65 \times 10^6$

As there was uncertainty as to the precise accuracy of the measurements of potential, it appeared best to remeasure the potentials with no current through the armature with the Siemens voltmeter placed as in the last experiment. Each machine was therefore run on open circuit with its magnets excited, and its potential was measured.

	Potential on Magnets in volts	Potential on Brushes	Speed per Minute	Potential at 880 Revs.
No. 1	25—25	90—90	880	90·0
No. 2	28—28	79—80	715—710	98·2

From which, since the formula is reduced to

$$I = \frac{A_2}{2l_2}(4\pi nc - 4\lambda mC),$$

the characteristic being practically straight, we infer:—

	Potential on Magnets	Potential on Brushes	Induction, $I = F(4\pi nc)$
No. 1	24	86·4	$2·82 \times 10^6$
No. 2	29	101·7	$3·30 \times 10^6$

We have further:—

$\lambda = 0·45$ for No. 1; $\qquad \lambda = 0·5$ for No. 2;

$\dfrac{4mC}{\nu} = 2,920;$ $\qquad \dfrac{\nu - 1}{\nu} 4mC \dfrac{A_2}{2l_2} = 443,800.$

	$\dfrac{4\lambda mC}{\nu}$	$\dfrac{\nu-1}{\nu}4\lambda mC\dfrac{A_2}{2l_3}$	$4\pi nc-\dfrac{4\lambda mC}{\nu}$	$F\left(4\pi nc-\dfrac{4\lambda mC}{\nu}\right)$	$F\left(4\pi nc-\dfrac{4\lambda mC}{\nu}\right)$ $-\dfrac{\nu-1}{\nu}4\lambda mC\dfrac{A_2}{2l_2}$
No. 1	1,314	199,700	7,586	$2\cdot41\times10^6$	$2\cdot21\times10^6$
No. 2	1,460	221,900	9,290	$2\cdot90\times10^6$	$2\cdot68\times10^6$

It has already appeared that experiment gives for I in No. 1 $2\cdot3\times10^6$, and in No. 2 $2\cdot65\times10^6$. The difference is probably due to error in estimating the lead of the brushes, which is difficult, owing to uncertainty in the position of the neutral line on open circuit.

8.

ON THE THEORY OF ALTERNATING CURRENTS, PARTICULARLY IN REFERENCE TO TWO ALTERNATE-CURRENT MACHINES CONNECTED TO THE SAME CIRCUIT.

[From the *Proceedings of the Institution of Electrical Engineers*, pp. 3—21. Nov. 13, 1884.]

IN my lecture on Electric Lighting, delivered before the Institution of Civil Engineers last year*, I considered the question of two alternate-current dynamo machines connected to the same circuit, but having no rigid mechanical connection between them, and I showed that, if two such machines be coupled in series, they will tend to nullify each other's effect; if parallel, to add their effects†. The subject is one which already has practical importance and application, and may have much more in the future; it is also one suited for discussion, and upon which discussion is desirable. I therefore venture to bring before the Society what I said in my lecture, some other ways of looking at the same subject, and an experimental verification,

---

* This Paper is reprinted on pages 57 to 83 of this volume.

† *22nd November*, 1884.—My attention has only to-day been called to a paper by Mr Wilde, published by the Literary and Philosophical Society of Manchester, December 15, 1868, also *Philosophical Magazine*, January, 1869. Mr Wilde fully describes observations of the synchronising control between two or more alternate-current machines connected together. I am sorry I did not know of his observations when I lectured before the Institution of Civil Engineers, that I might have given him the honour which was his due. If his paper had been known to those who have lately been working to produce large alternate-current machines, it would have saved them both labour and money.

together with solutions of other problems requiring similar treatment.

The general explanation amounting to proof, so far as machines in series are concerned, is given in the following extract from my lecture:—

"There remains one point of great practical interest in connection with alternate-current machines, How will they behave when two or more are coupled together, to aid each other in doing the same work? With galvanic batteries, we know very well how to couple them, either in parallel circuit or in series, so that they shall aid, and not oppose the effects of each other; but with alternate-current machines, independently driven, it is not quite obvious what the result will be, for the polarity of each machine is constantly changing. Will two machines coupled together run independently of each other, or will one control the movement of the other in such wise that they settle down to conspire to produce the same effect, or will it be into mutual opposition? It is obvious that a great deal turns upon the answer to this question, for in the general distribution of electric light it will be desirable to be able to supply the system of conductors from which the consumers draw by separate machines, which can be thrown in and out at pleasure. Now I know it is a common impression that alternate-current machines cannot be worked together, and that it is almost a necessity to have one enormous machine to supply all the consumers drawing from one system of conductors. Let us see how the matter stands. Consider two machines independently driven, so as to have approximately the same periodic time and the same electromotive force. If these two machines are to be worked together, they may be connected in one of two ways: they may be in parallel circuit with regard to the external conductor, as shown by the full line in Fig. 1, that is, their currents may be added algebraically and sent to the external circuit; or they may be coupled in series, as shown by the dotted line, that is, the whole current may pass successively through the two machines, and the electromotive force of the two machines may be added, instead of their currents. The latter case is simpler. Let us consider it first. I am going to show that if you couple two such alternate-current machines in series, they will so control each other's phase as to nullify each other, and that you will get no effect from them; and, as a corollary from that, I am going to show that if you couple them in parallel circuit, they will work perfectly well together, and the currents they produce will be

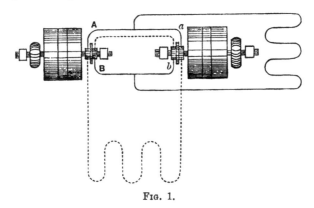

Fig. 1.

added; in fact, that you cannot drive alternate-current machines tandem, but that you may drive them as a pair, or, indeed, any number abreast. In diagram, Fig. 2, the horizontal line of abscissæ represents the time advancing from left to right; the full curves represent the electromotive forces of the two machines

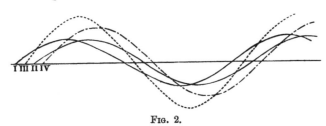

Fig. 2.

not supposed to be in the same phase. We want to see whether they will tend to get into the same phase or to get into opposite phases. Now, if the machines are coupled in series, the resultant electromotive force on the circuit will be the sum of the electromotive forces of the two machines. This resultant electromotive force is represented by the broken curve III; by what we have already seen in Formula IV (p. 65), the phase of the current must lag behind the phase of the electromotive force, as is shown in the diagram by curve IV, thus ———.———.———. Now, the work done in any machine is represented by the sum of the products of the currents and of the electromotive forces, and it is clear that as the phase of the current is more near to the phase of the lagging machine *II* than to that of the leading machine *I*, the lagging machine must do more work in producing electricity than the leading machine; consequently its velocity will be retarded, and its retardation will go on until the two machines settle down into exactly opposite phases, when no current will pass. The moral therefore is, do not attempt to couple two independently driven alternate-current machines in series. Now for the corollary. *A, B*, Fig. 1, represent the two terminals of an alternate-current machine; *a, b* the two terminals of another machine independently driven. *A* and *a* are connected together, and *B* and *b*. So regarded, the two machines are in series, and we have just proved that they will exactly oppose each other's effects—that is, when *A* is positive, *a* will be positive also; when *A* is negative, *a* is also negative. Now, connecting *A* and *a* through the comparatively high resistance of the external circuit with *B* and *b*, the current passing through that circuit will not much disturb, if at all, the relations of the two machines. Hence, when *A* is positive, *a* will be positive, and when *A* is negative, *a* will be negative also; precisely the condition required that the two machines may work together to send a current into the external circuit. You may therefore, with confidence, attempt to run alternate-current machines in parallel circuit for the purpose of producing any external effect. I might easily show that the same applies to a larger number: hence, there is no more difficulty in feeding a system of conductors from a number of alternate-current machines than there is in feeding it from a number of continuous-current machines. A little care only is required that the machine shall be thrown in when it has attained something like its proper velocity. A further corollary is that alternate currents with alternate-current machines as motors may theoretically be used for the transmission of power*."

Although the proof of this corollary regarding motors is similar to what we have just been going through, it may be instructive to give it. In the accompanying diagrams, Figs. 3 and 4, the full

---

* "Of course, in applying these conclusions, it is necessary to remember that the machines only *tend* to control each other, and that the control of the motive power may be predominant, and *compel* the two or more machines to run at different speeds."

lines I and II represent the electromotive forces of the two machines (generator and receiver); the dotted line, curve III (. . . .), the resultant electromotive force; and the curve IV, the resulting current, each in terms of the time, as abscissæ. The only difference between the two diagrams is, that in Fig. 3

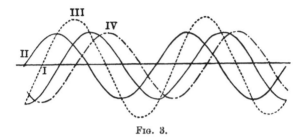

FIG. 3.

the two machines have equal electromotive forces, whilst in Fig. 4 *the receiving machine has double the electromotive force of the generator.* In both figures the receiving machine lags behind the phase of direct opposition to the generator by one quarter of

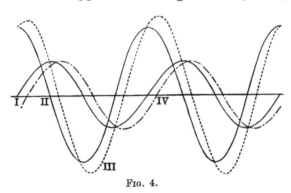

FIG. 4.

a period, or something less. Now observe, the resultant electromotive force must be in phase behind the receiver, but in advance of the generator. Also observe, the current must be in phase behind the resultant electromotive force, and may be one quarter of a period behind, provided only the self-induction be large enough compared with the resistance. The current will then be less than a quarter period behind the generator. This machine will do work upon the current, but the current will be more than a quarter period behind the receiving machine; therefore in the receiver the current does work upon the machine.

The subject is illustrated by the following problems. Of course any of them may be treated more generally by considering the machines as unequal, or by introducing other periodic terms, but I do not see that this would throw more light on the subject:—

I. *Two alternate-current machines, equal in all respects, are connected in series and independently driven at the same speed, to determine the current, etc., in each.*

Let γ be the coefficient of self-induction of each, r the resistance, x the current at time t, and $E \sin \frac{2\pi}{T}(t+\tau)$ and $E \sin \frac{2\pi}{T}(t-\tau)$ the electromotive forces. Then regarding the coefficient of self-induction as constant, which it is not exactly, and neglecting the effect of currents other than those in the copper wire, the equation of motion is

$$2\gamma x' + 2rx = E \left\{ \sin \frac{2\pi}{T}(t+\tau) + \sin \frac{2\pi}{T}(t-\tau) \right\},$$

or
$$\gamma x' + rx = E \sin \frac{2\pi t}{T} \cos \frac{2\pi \tau}{T};$$

whence
$$x = \frac{E \cos \frac{2\pi \tau}{T}}{r^2 + \left(\frac{2\pi\gamma}{T}\right)^2} \left\{ r \sin \frac{2\pi t}{T} - \frac{2\pi\gamma}{T} \cos \frac{2\pi t}{T} \right\}.$$

Work done by the leading machine per second

$$= \frac{E^2 \cos \frac{2\pi\tau}{T}}{2\left\{r^2 + \left(\frac{2\pi\gamma}{T}\right)^2\right\}} \cdot \left\{ r \cos \frac{2\pi\tau}{T} - \frac{2\pi\gamma}{T} \sin \frac{2\pi\tau}{T} \right\}$$

$$= \frac{E^2}{4\left\{r^2 + \left(\frac{2\pi\gamma}{T}\right)^2\right\}} \cdot \left\{ r\left(1 + \cos \frac{4\pi\tau}{T}\right) - \frac{2\pi\gamma}{T} \sin \frac{4\pi\tau}{T} \right\}.$$

From this at once follows that the leading machine does least work, and will tend to increase its lead until $\tau = \frac{T}{4}$, when the two machines will neutralise each other, as already proved geometrically. *The leading machine may actually become a motor and do* mechanical work, although its electromotive force is precisely equal to that of the following machine.

Considering the important case when r is negligible, we have

$$x = -\frac{E \cos \frac{2\pi\tau}{T} \cdot \cos \frac{2\pi t}{T}}{\frac{2\pi\gamma}{T}},$$

rate of working $= \dfrac{E^2 \sin \frac{4\pi\tau}{T}}{4 \cdot \frac{2\pi\gamma}{T}}.$

This is the maximum when $\tau = \dfrac{T}{8}$, and then it is equal to the maximum work which can be obtained from either machine when connected to a resistance only, which occurs when that resistance is $\dfrac{2\pi\gamma}{T}$; the current, also, is the same as when the maximum work is being done on resistance, and is $\dfrac{1}{\sqrt{2}}$ of the current the machine will give if short-circuited. The difference of potential between the two leads connecting the machines, whether $r = 0$ or not, is $E \cos \dfrac{2\pi t}{T} \sin \dfrac{2\pi\tau}{T}$. If there be no work done on the receiving machine and $r = 0$, $\tau = \dfrac{T}{4}$, and the amplitude of the difference of potential between the leads is E; if, on the other hand, the maximum work is being transmitted, the potential measured will be $\dfrac{1}{\sqrt{2}}$ of that observed when either machine is run on open circuit.

II. *Two machines are coupled parallel and connected to an external circuit resistance R.*

Let x_1, x_2 be currents in the two machines. The external current will be $x_1 + x_2$, and consequently the difference of potential at the junction, $R(x_1 + x_2)$.

Let the electromotive forces of the two machines regarded in this case as connected parallel be $E \sin \dfrac{2\pi (t \pm \tau)}{T}$ and let the self-induction and resistance of each be 2γ and $2r$.

The equations of motion then are

$$2\gamma x_1' + 2rx_1 = E \sin \frac{2\pi (t+\tau)}{T} - R(x_1 + x_2),$$

$$2\gamma x_2' + 2rx_2 = E \sin \frac{2\pi (t-\tau)}{T} - R(x_1 + x_2);$$

whence

$$\gamma(x_1' + x_2') + (R + r)(x_1 + x_2) = E \sin \frac{2\pi t}{T} \cdot \cos \frac{2\pi \tau}{T},$$

and

$$\gamma(x_1' - x_2') + r(x_1 - x_2) = E \cos \frac{2\pi t}{T} \sin \frac{2\pi \tau}{T}.$$

Solving these

$$x_1 + x_2 = \frac{E \cos \frac{2\pi \tau}{T}}{(r+R)^2 + \left(\frac{2\pi \gamma}{T}\right)^2} \left\{ (r+R) \sin \frac{2\pi t}{T} - \frac{2\pi \gamma}{T} \cos \frac{2\pi t}{T} \right\},$$

$$x_1 - x_2 = \frac{E \sin \frac{2\pi \tau}{T}}{r^2 + \left(\frac{2\pi \gamma}{T}\right)^2} \left\{ r \cos \frac{2\pi t}{T} + \frac{2\pi \gamma}{T} \sin \frac{2\pi t}{T} \right\}.$$

Electrical work done by the leading machine

$$= \tfrac{1}{2} E \sin \frac{2\pi (t+\tau)}{T} \{x_1 + x_2 + (x_1 - x_2)\}$$

$$= \tfrac{1}{4} \frac{E^2}{(r+R)^2 + \left(\frac{2\pi \gamma}{T}\right)^2} \left\{ (r+R) \cos^2 \frac{2\pi \tau}{T} - \frac{2\pi \gamma}{T} \sin \frac{2\pi \tau}{T} \cos \frac{2\pi \tau}{T} \right\}$$

$$+ \tfrac{1}{4} \frac{E^2}{r^2 + \left(\frac{2\pi \gamma}{T}\right)^2} \left\{ r \sin^2 \frac{2\pi \tau}{T} + \frac{2\pi \gamma}{T} \sin \frac{2\pi \tau}{T} \cos \frac{2\pi \tau}{T} \right\}.$$

This expression shows that *the leading machine does most work in all cases.* Suppose r is small compared with R and $\frac{2\pi \gamma}{T}$, also that $R = \frac{2\pi \gamma}{T}$, we have the work done per second

$$= \frac{E^2}{8R} \left\{ \cos^2 \frac{2\pi \tau}{T} + \sin \frac{2\pi \tau}{T} \cos \frac{2\pi \tau}{T} \right\}.$$

Make $\tau = -\dfrac{T}{8}$, and we see that the following machine will then do no work; when τ exceeds this, the following machine becomes a motor and absorbs electrical work.

III. *Suppose the terminals of an alternate-current machine are connected to a pair of conductors, the difference of potential between which is completely controlled by connection with other alternate-current machines.*

Let γ and R be the coefficient of self-induction and the resistance of the machine and its own conductors up to the point at which the potential is completely controlled. Let the difference of potential of the main conductors be $A \sin \dfrac{2\pi t}{T}$, and let the electromotive force of the machine be $B \sin \dfrac{2\pi(t-\tau)}{T}$.

Equation of motion is

$$\gamma x' + Rx = B \sin \frac{2\pi(t-\tau)}{T} - A \sin \frac{2\pi t}{T},$$

whence

$$x = \frac{1}{R^2 + \left(\frac{2\pi\gamma}{T}\right)^2} \left[B\left\{ R \sin \frac{2\pi(t-\tau)}{T} - \frac{2\pi\gamma}{T} \cos \frac{2\pi(t-\tau)}{T} \right\} \right.$$
$$\left. - A \left\{ R \sin \frac{2\pi t}{T} - \frac{2\pi\gamma}{T} \cos \frac{2\pi t}{T} \right\} \right].$$

Electrical work done by the machine in unit of time

$$= xB \sin \frac{2\pi(t-\tau)}{T}$$

$$= \frac{1}{R^2 + \left(\frac{2\pi\gamma}{T}\right)^2} \left[\frac{B^2 R}{2} - \frac{AB}{2} \left\{ R \cos \frac{2\pi\tau}{T} + \frac{2\pi\gamma}{T} \sin \frac{2\pi\tau}{T} \right\} \right].$$

If τ be positive, that is, if the machine be lagging in its phase, work done is less than if it be negative; hence τ will tend to zero, or the machine will tend to adjust itself to add its currents to that of the system of conductors. *The machine may act as a motor even though its electromotive force be greater than that of the system*, for let

$$\frac{R}{\frac{2\pi\gamma}{T}} = \tan \frac{2\pi\phi}{T},$$

work (electric) done by machine

$$= \frac{B^2 R}{2\left\{R^2 + \left(\frac{2\pi\gamma}{T}\right)^2\right\}} - \frac{AB}{2\left\{R^2 + \left(\frac{2\pi\gamma}{T}\right)^2\right\}^{\frac{1}{2}}} \sin\frac{2\pi(\phi+\tau)}{T};$$

this has a minimum value when $\phi + \tau = \dfrac{T}{4}$, and then the mechanical work done by machine or electrical work received by the machine

$$= \frac{B}{2\left\{R^2 + \left(\frac{2\pi\gamma}{T}\right)^2\right\}^{\frac{1}{2}}} \left(A - \frac{RB}{\left\{R^2 + \left(\frac{2\pi\gamma}{T}\right)^2\right\}^{\frac{1}{2}}}\right),$$

and this is positive provided

$$\frac{A}{B} > \frac{R}{\left\{R^2 + \left(\frac{2\pi\gamma}{T}\right)^2\right\}^{\frac{1}{2}}}.$$

There are two or three other problems of sufficient interest to make it worth while giving them here, although not directly relating to alternate-current machines coupled together.

IV. *To determine the law of an alternate current through an electric arc.*

It has been shown by Joubert that in an arc the difference of potential is of approximately constant numerical value, reversing its value discontinuously with the reversal of the current, probably at the instant of reversal of current. We shall assume, then, that there is in the arc a constant electromotive force, A, always opposed to the current, except when the current ceases, and that then its value is zero.

The equation of motion is

$$\gamma x' + Rx = E\sin\frac{2\pi t}{T} \mp A,$$

the negative sign being taken when x is positive, the positive when x is negative. Solving generally

$$x = \mp\frac{A}{R} + \frac{E}{\left(\frac{2\pi\gamma}{T}\right)^2 + R^2}\left(-\frac{2\pi\gamma}{T}\cos\frac{2\pi t}{T} + R\sin\frac{2\pi t}{T}\right) + Ce^{-\frac{R}{\gamma}t},$$

ON THE THEORY OF ALTERNATING CURRENTS.

this equation will continuously hold good for a half period from $x = 0$ to $x = 0$ again, but at each half period the arbitrary constant C is changed with the sudden change of sign of A. It is determined by the consideration that, if for a certain value t_0 of t, x should vanish, it shall vanish again when $t = t_0 + \dfrac{T}{2}$. This applies to the case when E is sufficiently large, as is practically the case, but if the current should cease for a finite time this condition will be varied, and instead of it we have the condition $x = 0$ when $E \sin \dfrac{2\pi t}{T} = A$. This latter case I do not propose to consider further.

Let $\dfrac{2\pi \gamma}{RT} = \tan \dfrac{2\pi t_1}{T}$;

$$x = -\frac{A^*}{R} + \frac{E}{\sqrt{\left\{R^2 + \left(\frac{2\pi \gamma}{T}\right)^2\right\}}} \sin \frac{2\pi (t - t_1)}{T} + Ce^{-\frac{R}{\gamma}t}.$$

Putting $t = t_0$ and $t = t_0 + \dfrac{T}{2}$ we have

$$0 = -\frac{A}{R} + \frac{E}{\sqrt{\left\{R^2 + \left(\frac{2\pi \gamma}{T}\right)^2\right\}}} \sin \frac{2\pi (t_0 - t_1)}{T} + Ce^{-\frac{R}{\gamma}t_0},$$

$$0 = -\frac{A}{R} - \frac{E}{\sqrt{\left\{R^2 + \left(\frac{2\pi \gamma}{T}\right)^2\right\}}} \sin \frac{2\pi (t_0 - t_1)}{T}$$
$$+ Ce^{-\frac{R}{\gamma}t_0} \cdot e^{-\frac{RT}{2\gamma}},$$

equations to determine t_0 and C.

Eliminating C,

$$\frac{RE}{A \sqrt{\left\{R^2 + \left(\frac{2\pi \gamma}{T}\right)^2\right\}}} \cdot \sin \frac{2\pi (t_0 - t_1)}{T} = -\tanh \frac{RT}{4\gamma}.$$

Having obtained t_0, C is given by equation

$$\frac{2A}{R} = Ce^{-\frac{R}{\gamma}t_0}\left(1 + e^{-\frac{R}{2\gamma}T}\right).$$

This gives the complete solution of the problem.

---

* That is, taking a half-period in which x is positive and the negative sign affixed to A. [ED.]

A case of special importance is that in which R is small; let us therefore consider the case $R = 0$, the solution then is

$$\gamma x = -\frac{T}{2\pi} E \cos \frac{2\pi t}{T} - At + C.$$

In the same way as before

$$E \cos \frac{2\pi t_0}{T} = \frac{A\pi}{2},$$

$$C = A \left(t_0 + \frac{T}{4} \right).$$

The limiting case to which the solution applies is given by $x = 0$ when $t = t_0 + \frac{T}{2}$.

$$E \sin \frac{2\pi t_0}{T} = A,$$

whence

$$E^2 - A^2 \left(1 + \frac{\pi^2}{4} \right),$$

or

$$A = E \times 0\cdot538.$$

Roughly, we may say that, in order that the current may not cease for a finite time, E must be at least double of A; A will of course depend upon the length of the arc. The work done in the arc will be proportional to the arithmetical mean value of the current taken without regard to sign. This is of course quite a different thing from the mean current as measured by an electrodynamometer. Let us examine what error is caused by estimating the work done in the arc as equal to the current measured by the dynamometer multiplied by the mean difference of potential.

The actual work done per second

$$= \frac{2A}{T} \int_{t_0}^{t_0+\frac{T}{2}} x dt$$

$$= \frac{2A}{\pi\gamma} \cdot \frac{T}{2\pi} \cdot \sqrt{E^2 - \frac{\pi^2 A^2}{4}}.$$

The mean square of the current, as measured by the electrodynamometer, is

$$\frac{2}{T} \int_{t_0}^{t_0+\frac{T}{2}} x^2 dt = \overline{\left.\frac{T}{2\pi\gamma}\right|^2} \left\{ \frac{E^2}{2} - 2A^2 \right\} + \frac{A^2 T^2}{48\gamma^2},$$

and the work done by this current is apparently the square root of the above expression multiplied by A. It is easy to see that this is greater in all cases than the work done, but it is worth while to examine the extent of the error. If we treated the arc as an ordinary resistance, we should assume work per second

$$= \frac{A}{\gamma}\sqrt{\left(\frac{T}{2\pi}\right)^2\left(\frac{E^2}{2} - 2A^2\right) + \frac{A^2T^2}{48}}.$$

Taking a fairly practical case, assume $A = \frac{2}{5}E$, we have actual work per second

$$= \frac{A^2T}{\gamma} \cdot \frac{1}{\pi^2}\sqrt{\frac{25}{4} - \frac{\pi^2}{4}}$$

$$= \frac{A^2}{\gamma} T \frac{\sqrt{15}}{20} \text{ nearly,}$$

work done estimated by electro-dynamometer

$$= \frac{A^2T}{\gamma}\sqrt{\frac{1}{40}\left(\frac{1}{2}\frac{25}{4} - 2\right) + \frac{1}{48}} \text{ nearly}$$

$$= \frac{A^2T}{\gamma}\frac{20}{1}\sqrt{\frac{235}{12}},$$

or nearly $\frac{1}{8}$ part too much. This will suffice to show that the matter is not a mere theoretical refinement. Another erroneous method of estimating the power developed in an arc is, to replace by a resistance and adjust this resistance till the current, as measured by an electro-dynamometer, is the same as with the arc, and assume that the work done in the resistance is the same as the work done in the arc.

Returning to the expression

$$\frac{2A}{\pi\gamma} \cdot \frac{T}{2\pi}\sqrt{E^2 - \frac{\pi^2 A^2}{4}}$$

we may enquire, given TA and the dimensions of the machine, how ought it to be wound or its coils connected, that most work may be done in the arc? If the number of convolutions be varied, E will vary as the convolutions, γ as their square, therefore $\gamma \propto E^2$; we are therefore to determine E so that $\frac{1}{E^2}\sqrt{E^2 - \frac{\pi^2 A^2}{4}}$ is a maximum which occurs when $E = \pi A$. When the resistance of the circuit is taken into account this result will be modified. It suffices to prove that it is desirable that the potential of the machine should be materially in excess of that required to maintain the arc.

V.* In all that precedes it is assumed, not only that γ is constant, but that the copper conductor of the armature is the only conductor moving in the field. If there be iron cores in the armature, we shall approximate to the effect by regarding such cores as a second conducting circuit. Slightly changing the notation, let L be coefficient of self-induction of the copper circuit, N coefficient of self-induction of the iron circuit and R' its resistance, I' the magnetic induction of the field magnets upon the iron circuit and M the coefficient of mutual induction of the two circuits, y the current in the iron. The equations of motion are obtained from the expression for the energy, viz.

and are
$$\tfrac{1}{2}\{Lx^2 + 2Mxy + Ny^2 - 2Ix - 2I'y\},$$

$$Lx' + My' + Rx = \frac{dI}{dt} = \frac{2\pi A}{T} \cos \frac{2\pi t}{T},$$

$$Mx' + Ny' + R'y = \frac{dI'}{dt} = \frac{2\pi B}{T} \cos \frac{2\pi t}{T},$$

for in general the iron cores and the copper conductor are symmetrically arranged. Assume

$$x = a \sin \frac{2\pi t}{T} + b \cos \frac{2\pi t}{T},$$

$$y = a' \sin \frac{2\pi t}{T} + b' \cos \frac{2\pi t}{T},$$

and substitute in the equations of motion, we have the following four equations to determine the constants, a, b, a', b',

$$a \frac{2\pi L}{T} + a' \frac{2\pi M}{T} + Rb = \frac{2\pi A}{T},$$

or

and

$$\left.\begin{array}{l} aL + a'M + b\dfrac{TR}{2\pi} = A \\[4pt] bL + b'M - a\dfrac{TR}{2\pi} = 0 \\[4pt] aM + a'N + b'\dfrac{TR'}{2\pi} = B \\[4pt] bM + b'N - a'\dfrac{TR'}{2\pi} = 0 \end{array}\right\}.$$

* *Vide* also *Encyclopædia Britannica*, article "Lighting."

ON THE THEORY OF ALTERNATING CURRENTS. 147

These equations contain the solution of the problem, but are too cumbersome to be worth while solving generally; we will however prove the statements made in the lecture before the Civil Engineers.

1st. Compare short circuit and open circuit, that is $R = 0$, very nearly, and $R = \infty$'. In the former case we find that work done in the iron is diminished, and if $B = \dfrac{AM}{L}$ we have the paradoxical result that there are no currents induced in the iron of the cores and no work is required to drive the machine. This of course can never actually occur, because R can never absolutely vanish. It suffices to show, however, that the current in the copper circuit may diminish the whole power required to drive the machine, to an amount less than the power required to drive the machine on open circuit.

2nd. The other statement related to the effect of the currents in the iron upon the currents produced in the copper circuit. Assume that the effect is a small one, for a first approximation. Neglect it, that is, treat the currents in the iron and the currents in the copper as independent of each other, and then see how each would disturb the other.

The first approximation then is

$$\left\{ \begin{array}{l} a = \dfrac{AL}{L^2 + \dfrac{T^2 R^2}{4\pi^2}}; \quad a' = \dfrac{BN}{N^2 + \dfrac{T^2 R'^2}{4\pi^2}}; \\[2ex] b = \dfrac{A \dfrac{TR}{2\pi}}{L^2 + \dfrac{T^2 R^2}{4\pi^2}}; \quad b' = \dfrac{B \dfrac{TR'}{2\pi}}{N^2 + \dfrac{T^2 R'^2}{4\pi^2}}. \end{array} \right.$$

If we substitute these in the general equations as corrections, we have

$$\left\{ \begin{array}{l} aL + b\dfrac{TR}{2\pi} = A - \dfrac{BNM}{N^2 + \dfrac{T^2 R'^2}{4\pi^2}} \\[2ex] -a\dfrac{TR}{2\pi} + bL = -\dfrac{B\dfrac{TR'}{2\pi}M}{N^2 + \dfrac{T^2 R'^2}{4\pi^2}}, \end{array} \right.$$

10—2

which shows that the disturbing effect of each circuit upon the other is to diminish the apparent electromotive force, and to accelerate its phase.

VI. A very similar problem is that of secondary generators or induction coils, whether used for the conversion of high potentials to low, or the reverse. To treat it generally, taking the magnetisation of the iron cores, which are always used, as a non-linear function of the currents in the coils, would be a matter of much difficulty; we therefore assume, as is usual, that the coefficients of induction are constants, noting in passing that this is not strictly the fact, though it is very nearly the fact, when the cores are not saturated, and when the lines of magnetic induction pass through non-magnetic space.

Let, then, R, r be the resistances of the primary and secondary circuits,

L coefficient of self-induction of the primary,

N coefficient of self-induction of the secondary,

M coefficient of mutual induction of the two circuits,

x and y the currents in the two circuits at time t,

X the electromotive force applied in the primary circuit by an alternate-current dynamo machine or otherwise,

the equations of motion will be

$$\left. \begin{array}{l} Lx' + My' + Rx = X \\ Mx' + Ny' + ry = 0 \end{array} \right\}.$$

Various assumptions may be made as to X, but that most likely to be adopted in the practical work of secondary generators is that X is kept so adjusted that

$$x = A \cos nt, \text{ where } n = \frac{2\pi}{T},$$

and to enquire how X will depend on the resistances.

$$Ny' + ry = nAM \sin nt,$$

$$y = \frac{nAM}{n^2N^2 + r^2}(-nN \cos nt + r \sin nt),$$

$$X = A\left[\left\{-nL + \frac{n^3NM^2}{n^2N^2 + r^2}\right\} \sin nt + \left\{\frac{M^2n^2r}{n^2N^2 + r^2} + R\right\} \cos nt\right].$$

As in the case of the dynamo machine, the work done in the secondary circuit is greatest when $r = nN$. The expression for

ON THE THEORY OF ALTERNATING CURRENTS. 149

X serves to show that when the secondary is short-circuited a *lower* electromotive force of the generating circuit is required than when it is on open circuit. In induction coils the electrostatic capacity of the coils themselves has important effects. An illustration of the effect of electrostatic induction is found in the old-fashioned Ruhmkorff coils. These were not wound symmetrically, but in such wise that one end of the secondary coil was on the whole towards the inside, the other towards the outside of the bobbin. In such coils a spark to earth may be obtained from the outside end, but not from the inside. The reason is that the outer convolutions have smaller electrostatic capacity than the inner ones. The terminals may be made to give equal sparks by the simple expedient of laying a piece of tinfoil around the whole coil and connecting it to earth.

VII. Some time ago Dr Muirhead told me that he had observed that the effect of an alternate-current machine could be increased by connecting it to a condenser. This is not difficult to explain: it is a case of resonance analogous to those which are so familiar in the theory of sound and in many other branches of physics.

Take the simplest case, though some others are almost as easy to treat. Imagine an alternate-current machine with its terminals connected to a condenser; it is required to find the amplitude of oscillation of potential between the two sides of the condenser. Let $R\gamma$ be the resistance and self-induction of the machine, $E \sin \dfrac{2\pi t}{T}$ its electromotive force, C the capacity of the condenser, V the difference of potential sought, and x the current in the machine, then

$$CV = x$$

and
$$\gamma x' + Rx = E \sin \frac{2\pi t}{T} - V,$$

whence
$$\gamma x'' + Rx' = \frac{2\pi}{T} E \cos \frac{2\pi t}{T} - \frac{x}{C},$$

$$x = \frac{\left\{1 - C\gamma \dfrac{2\pi^2}{T}\right\} \cos \dfrac{2\pi t}{T} + RC \dfrac{2\pi}{T} \sin \dfrac{2\pi t}{T}}{\left\{1 - C\gamma \left(\dfrac{2\pi}{T}\right)^2\right\}^2 + R^2 C^2 \left(\dfrac{2\pi}{T}\right)^2} \cdot \frac{2\pi EC}{T},$$

$$V = \frac{\left\{1 - C\gamma\left(\frac{2\pi}{T}\right)^2\right\}\sin\frac{2\pi t}{T} - RC\frac{2\pi}{T}\cos\frac{2\pi t}{T}}{\left\{1 - C\gamma\left(\frac{2\pi}{T}\right)^2\right\}^2 + R^2C^2\left(\frac{2\pi}{T}\right)^2} \cdot E;$$

amplitude of V is therefore

$$= \frac{1}{\sqrt{\left\{1 - C\gamma\left(\frac{2\pi}{T}\right)^2\right\}^2 + R^2C^2\left(\frac{2\pi}{T}\right)^2}} \cdot E.$$

Now suppose $E = 100$ volts, the machine would light up an incandescent lamp of about 69 volts; let $T = \frac{1}{200}$ second, $C = 100$ microfarads, and $\frac{2\pi\gamma}{T} =$ eight ohms, and $R = \frac{1}{10}$ ohm, all figures which could be practically realised, we have amplitude of $V = 80\,E$ roughly, or the apparent electromotive force would be increased eighty-fold.

We now return to the principal subject of the present communication. Some attempts have been made to verify the proposition that two alternate-current machines can be advantageously connected parallel, but I believe, till recently, without success. I had no convenient opportunity for testing the point myself till last summer, when I had two machines of De Meritens, intended for the lighthouse of Tino*, in my hands. I have made no determinations of the constants of these machines, but between three and four years ago I thoroughly tested a pair of similar machines now in use at a lighthouse in New South Wales*. Each machine has five rings of sixteen sections, and 40 permanent magnets. The resistance of the whole machine as connected for lighthouse work (a single arc) was 0·0313, its electromotive force (E) when running 830 revolutions per minute, 95 volts and $\left(\frac{2\pi\gamma}{T}\right) = 0·044$ ohms. It was further remarked that the loss of power was least with a maximum load, as is shown in the following table:—

Power applied as measured in belt	3·1	4·8	5·6	6·5	5·4
Electric power developed............	0·7	3·4	4·3	5·7	3·4
Mean current in ampères	7·7	38·6	51·7	73·6	151

* The engines, dynamos, lamps, optical apparatus, and lanterns of both these lighthouses have been supplied by Messrs Chance Bros. & Co.

ON THE THEORY OF ALTERNATING CURRENTS. 151

This result illustrates well the conclusion arrived at in Problem V above.

Last summer the two machines for Tino were driven from the same countershaft by link bands, at a speed of 850 to 900 revolutions per minute; the pulleys on the countershaft were sensibly equal in diameter, but those on the machines differed by rather more than a millimetre, one being 300, the other 299 mm. in diameter (about); thus the two machines had not when unconnected exactly the same speed. The pulleys have since been equalised. The bands were of course put on as slack as practicable, but no special appliance for adjusting the tightness of the bands was used. The experiment succeeded perfectly at the very first attempt. The two machines, being at rest, were coupled in series with a pilot incandescent lamp across the terminals, the two bands were then simultaneously thrown on: for some seconds the machines almost pulled up the engine. As the speed began to increase, the lamp lit up intermittently, but in a few seconds more the machines dropped into step together, and the pilot lamp lit up to full brightness and became perfectly steady and remained so. An arc lamp was then introduced, and a perfectly steady current of over 200 ampères drawn off without disturbing the harmony. The arc lamp being removed, a Siemens electrodynamometer was introduced between the machines, and it was found that the current passing was only 18 ampères, whereas, if the machines had been in phase to send the current in the same direction, it would have been more than ten times as great. On throwing off the two bands simultaneously, the machines continued to run by their own momentum, with retarded velocity. It was observed that the current, instead of diminishing from diminished electromotive force, steadily increased to about 50 ampères, owing to the diminished electrical control between the machines, and then dropped off to zero as the machines stopped. Professor Adams will, I hope, give an account of experiments he has tried with me, and on other occasions, at the South Foreland. With De Meritens' machines, I regard coupling two or more machines parallel as practically the best way of obtaining exceptionally great currents when required in a lighthouse for penetrating a thick atmosphere.

9.

NOTE ON THE THEORY OF THE ALTERNATE CURRENT DYNAMO.

[From the *Proceedings of the Royal Society*, Vol. XLII., pp. 167—170.]

Received February 17, 1887.

ACCORDING to the accepted theory of the alternate-current dynamo, the equation of electric current in the armature is $\gamma \dot{y} + Ry = $ periodic function of t, where γ is a constant coefficient of self-induction. This equation is not strictly true, inasmuch as γ is not in general constant*, but it is a most useful approximation. My present purpose is to indicate how the values of γ and of the periodic function representing the electromotive force can be calculated in a machine of given configuration.

To fix ideas, we will suppose the machine considered to have its magnet cores arranged parallel to the axis of rotation, that the cores are of uniform section, also that the armature bobbins have iron cores, so that we regard all the lines of induction as passing either through an armature coil, or else between adjacent poles entirely outside the armature. The sketch shows a development of the machine considered. The iron is supposed to be so arranged that the currents induced therein may be neglected. We further suppose for simplicity that the line integral of magnetic force within the armature core may be neglected.

Let A_1 be the effective area of the space between the pole piece and armature core when the cores are in line, l_1 the distance from iron to iron.

---

* "On the Theory of Alternating Currents," *Telegr. Engin. Journ.* vol. XIII, 1884, p. 496.

Let A_2 be the section of magnet core, l_2 the effective length of a pair of magnet limbs, so that l_2 may be regarded as the length of the lines of force as measured from one pole face to the next.

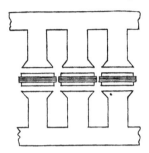

Let m be the number of convolutions in a pair of magnet limbs, and

n, the convolutions in one armature section,

T, the periodic time.

The time is measured from an epoch when the armature coil we shall consider is in a symmetrical position in a field which we shall regard as positive.

x and y are the currents in the magnet and armature coils, the positive direction being that which produces the positive field at time zero.

At time t the armature coil considered has area A_1',
$$= b_0 + b_1 \cos(2\pi t/T) + b_2 \cos(4\pi t/T) + \&c.$$
in a positive field; and area A_1'',
$$= b_0 - b_1 \cos(2\pi t/T) + b_2 \cos(4\pi t/T) - \&c.$$
in a negative field, where
$$b_0 + b_1 + b_2 + \ldots = A_1,$$
and
$$b_0 - b_1 + b_2 + \ldots = 0.$$

The coefficients b_0, b_1, &c., are deducible by Fourier's theorem from a drawing of the machine under consideration.

Let I be the total induction in the magnet core, and let, at time t, I be distributed into I' through A_1', I'' through A_1'' and I''' as a waste field to the neighbouring poles.

The line integral of magnetic force from the pole to either adjacent pole is I'''/k, where k is a constant.

We have first to determine I', I'', I''', in terms of x and y.

Take the line integral of magnetic force in three ways through the magnets, and respectively through area A_1', through area A_1'', and across between the adjacent poles—

$$l_2 f\left(\frac{I}{A_2}\right) + 2l_1 \frac{I'}{A_1'} = 4\pi mx + 4\pi ny,$$

$$l_2 f\left(\frac{I}{A_2}\right) + 2l_1 \frac{I''}{A_1''} = 4\pi mx - 4\pi ny,$$

$$l_2 f\left(\frac{I}{A_2}\right) + \frac{I'''}{k} = 4\pi mx\,;$$

whence $\qquad I' + I'' + I''' = I$

$$= \left(\frac{A_1' + A_1''}{2l_1}\right)\left(4\pi mx - l_2 f\left(\frac{I}{A_2}\right)\right) + \frac{A_1' - A_1''}{2l_1} \cdot 4\pi ny.$$

When t, x, and y are given, this would suffice to determine I by means of the known properties of the material of the magnets as represented by the function f. We will, however, consider two extreme cases between which other cases will lie.

First.—Suppose that the intensity of induction in the magnet cores is small, so that $l_2 f(I/A_2)$ may be neglected, the iron being very far from saturation. We have—

$$I' - I'' = \frac{4\pi}{2l_1}\left\{m(A_1' - A_1'')x + n(A_1' + A_1'')y\right\}$$

$$= \frac{4\pi}{l_1}\left\{m\left(b_1 \cos\frac{2\pi t}{T} + b_3 \cos\frac{6\pi t}{T} + \ldots\right)x\right.$$

$$\left. + n\left(b_0 + b_2 \cos\frac{4\pi t}{T} + \ldots\right)y\right\}.$$

We see that the coefficient of self-induction γ in general contains terms in $\cos(4\pi t/T)$.

Second.—In actual work it would be nearer the truth to suppose that the magnetising current x is so great that the induction I may be regarded as constant, and the quantity $l_2 f(I/A_2)$ as considerable. But as small changes in I imply very great changes

ON THE THEORY OF THE ALTERNATE CURRENT DYNAMO. 155

in $l_2 f(I/A_2)$, its value cannot be regarded as known. We have then—

$$2l_1 \frac{I'}{A_1'} = \frac{2l_1}{A_1' + A_1'' + 2kl_1} \left(I - \frac{A_1' - A_1''}{2l_1} \cdot 4\pi m y \right) + 4\pi m y,$$

$$2l_1 \frac{I''}{A_1''} = \frac{2l_1}{A_1' + A_1'' + 2kl_1} \left(I - \frac{A_1' - A_1''}{2l_1} \cdot 4\pi m y \right) - 4\pi n y,$$

whence

$$I' - I'' = \frac{(A_1' - A_1'') I}{A_1' + A_1'' + 2kl_1} - \left\{ \frac{(A_1' - A_1'')^2}{A_1' + A_1'' + 2kl_1} - (A_1' + A_1'') \right\} \frac{4\pi n y}{2l_1}$$

$$= \frac{(A_1' - A_1'') I}{A_1' + A_1'' + 2kl_1} + \frac{4A_1' A_1'' + 2kl_1 (A_1' + A_1'')}{A_1' + A_1'' + 2kl_1} \cdot \frac{4\pi n y}{2l_1}.$$

For illustration, consider the simplest possible case: let $b_0 = b_1 = \tfrac{1}{2} A_1$, and $b_2 = b_3 = \ldots = 0$, and let $2kl_1$ be negligible; we have—

$$I' - I'' = I \cos \frac{2\pi t}{T} + A_1 \sin^2 \frac{2\pi t}{T} \cdot \frac{4\pi n y}{2l_1},$$

and the equation of current will be—

$$Ry = n \left\{ \frac{2\pi}{T} I \sin \frac{2\pi t}{T} - \frac{2\pi n A_1}{l_1} \frac{d}{dt} \left(\sin^2 \frac{2\pi t}{T} \cdot y \right) \right\},$$

instead of the simple and familiar linear equation.

10.

ALTERNATE CURRENT DYNAMO-ELECTRIC MACHINES.

[From the *Philosophical Transactions of the Royal Society*, 1896, pp. 229—252.]

Received April 4,—Read May 2, 1895.

THIS paper deals experimentally with the current induced in the coils and in the cores of the magnets of alternate current machines by the varying currents and by the varying positions of the armature. It is shown that such currents exist, and that they have the effect of diminishing to a certain extent the electromotive force of the machine when it is working on resistances as a generator without having a corresponding effect upon the phase of the armature current. It is also shown that preventing variations in the coils of the electromagnet does not, in the machine experimented upon, greatly affect the result, and that the effect of introducing copper plates between the magnets and the armature has not a very great effect upon the electromotive force of the armature, the conclusion being that the conductivity of the iron cores is sufficient to produce the main part of the effect. A method of determining the efficiency of alternate current machines is illustrated and the results of the experiments for this determination are utilised to show that in certain cases of relation of phase of current to phase of electromotive force the effect of the local currents in the iron cores is to increase instead of to diminish the electromotive force of the machine.

* The large majority of the experiments herein described were made in the summer of 1893 and a considerable part of the paper was then written. We have to thank Mr F. Lydall, one of the student demonstrators at King's College at that time, for much assistance.

I.

In algebraic discussions of the theory of alternate current machines, it has usually been assumed that the electromotive force due to the magnets is a periodic function, the same whether there is a current in the armature or not, and that the effect of the current in the armature can be represented by regarding the armature as having self-induction. It has been pointed out, too, that the coefficient of self-induction will generally vary with the position of the armature in the field. To state exactly the same thing in another way it has been assumed that the electromotive force of the magnets is a periodic function independent of the current in the armature, and that the effect of the armature current on the induction through the armature can be represented as an armature reaction which vanishes at the moment when the current in the armature vanishes.

To state the matter in the form of an equation, let E be the electromotive force of the machine on open circuit, R the resistance of the armature circuit, x the current in the armature, T the periodic time, then it is assumed that

$$Rx = E - (Lx)^{\cdot};$$

E being independent of x, L being, if you please, a coefficient of self-induction constant or variable, or, if you prefer it, Lx representing the change in the induction through the armature due to the current in the armature, vanishing with x.

It is easy to see that this statement is true in some cases. For example, it is very nearly true in the older machines with permanent magnets. Or imagine a machine without iron in the magnets or armature, consisting merely of two circuits—one the magnet circuit, the other the armature circuit—moveable in relation to each other. If the current in the magnet circuit is kept precisely constant, either by inserting a great self-induction in its circuit external to the machine, or by inserting such a resistance and using so high an electromotive force that any disturbing electromotive forces are inappreciable compared with it, the preceding statement is strictly accurate. But if the magnet current is not forced to be constant the problem is more complicated.

Stating the matter in the language of self- and mutual-induction, let x and y be the currents in armature and magnet circuits,

R and r their resistances, L and N their self-induction, $M \sin 2\pi t/T$ their mutual-induction, F the constant electromotive force applied to the magnet, the equations for the system are:—

$$Rx = -M(y \sin 2\pi t/T)\dot{} - L\dot{x}$$
$$ry = F - M(x \sin 2\pi t/T)\dot{} - N\dot{y}$$

These equations can be solved by approximations if the variations in the value of y are small.

First,

$$y = \frac{F}{r}; \quad x = -\frac{MF}{r}\frac{2\pi}{T}\frac{R \cos 2\pi t/T + 2\pi L/T \sin 2\pi t/T}{R^2 + (2\pi L/T)^2}.$$

Second, substituting this value of x, we obtain

$$ry = F + \frac{M^2 F}{r} \cdot \frac{2\pi/T}{R^2 + (2\pi L/T)^2}$$
$$\left\{\frac{R}{2} \sin \frac{4\pi t}{T} + \frac{\pi L}{T}\left(1 - \cos \frac{4\pi t}{T}\right)\right\}\dot{} - N\dot{y}.$$

This gives periodic terms in y, the period being one-half the period of the mutual induction.

Introducing these terms into the first equation, we see that the term in $2\pi t/T$ in the electromotive force of the machine is modified, and that terms in $6\pi t/T$ are introduced. The former may have real practical importance, and it is one of the objects of the present paper to ascertain how far it exists and is of importance in actual machines.

Returning to machines as ordinarily constructed, in these the current in the magnet coils is not compelled to be constant, and any rapid variation of the induction in the magnet cores will induce currents in those cores. The variations in the current in the magnet coils and the currents in the cores both tend to annul the variations in the induction in the core arising from the current in the armature, but they will not tend to alter the *average* effect of the currents in the armature on the induction through the magnets. That there will be such an average effect is not difficult to see. Suppose the armature coils to be fixed in line with the magnets of the machine, any current in the armature will have its full effect in increasing or diminishing the field through the magnets. Suppose the armature coils to be fixed midway between the magnets, any current in the armature will then have practically little or no effect in increasing or diminishing the field through the magnets. If the armature be connected through resistance, inductive or non-inductive, and the machine run in the ordinary way, the current in the armature will lag in phase behind the electromotive force E. The result is that when the armature is opposite to the magnets there is a current in the armature tending to demagnetise the magnets, and adding together the effects of the armature in all positions, there is an average effect tending to demagnetise the magnets. If the machine had a constant current round the magnets and divided iron in the magnets, this average effect, as well as its variations, would be fully accounted for by the term $(Lx)'$, call it self-induction or call it armature reaction, as you please. But inasmuch as the variations are in part annulled by the variations of current in the magnet-winding and the local currents in the magnet cores, we have a part of the diminution of the electromotive force E of the machine unaccompanied by retardation of phase of the current in the armature.

Before giving any experimental results, it will be well to describe the machines used and the general method of experiment adopted.

The two dynamos experimented upon were constructed by Messrs Siemens Bros., and are of the same pattern and size but are of an old type. They are mounted upon a common base-plate, their pulleys being provided with flanges and bolts, so that any desired phase difference can be given to the armatures, for the accurate setting of which a graduated circle is provided, or so that the armatures can be run independently of each other. The pulleys have each a diameter of 12 inches, and are suited for a 6-inch belt—the shaft is prolonged for the purpose of carrying a revolving contact-maker and a small pulley for driving a Buss tachometer.

Each dynamo has a series of 24 electromagnets (see fig. 1), there being 12 on either side of the armature. The core of each magnet (A) is of wrought-iron $2\frac{1}{16}$ inches diameter, and $6\frac{1}{4}$ inches long: and is wound with 5 layers, 35 convolutions per layer, of copper wire 3·5 millims. diameter. The electromagnets are bolted to circular cast-iron frames (B), which serve also for supporting the bearings of the armature shaft. The centres of the 12 electromagnets on each frame are equally spaced out on a circle $8\frac{7}{8}$ inches radius concentric with the axle of the machine. Each cast-iron frame has a cross-sectional area of 4·8 sq. inches. The opposing pole pieces of the electromagnets have an air space of $1\frac{1}{8}$ inch

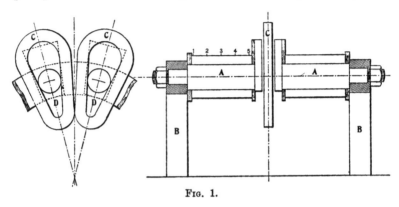

Fig. 1.

between them, through which the armature coils rotate. The 24 electromagnet windings are coupled in series, and have a total resistance of 1·8 ohms, the normal existing current being about 22 ampères.

The armature of each dynamo consists of 12 coils or bobbins (C) with wooden cores (D) $\frac{7}{8}$ inch thick. Each core is 7 inches long (radially), with rounded ends—the outer being struck to a circle $3\frac{1}{8}$ inches and the inner $1\frac{1}{8}$ inches diameter. The ends of the respective coils are brought to a screw-plug commutator board fixed to the shaft, by means of which a series of combinations can be made. Each coil consists of 10 layers, 8 convolutions per layer, 2·2 millims. copper wire, having a resistance (cold) of ·187 ohm, and at a speed of 1000 revolutions per minute, with normal excitation and fully loaded, is intended to give 50 volts at its extremities. The full load current for each coil is 16·6 ampères, so that at 1000 revolutions per minute, or a frequency of 100 complete periods per second, the machine should give, with its 12 armature coils in parallel, an output of 200 ampères 50 volts. The two terminal rings of the screw-plug commutator are connected by conductors to two gun-metal collector rings, insulated from one another and from the shaft by means of ebonite. Each collector ring is provided with two 1-inch copper-wire brushes carried by adjustable bar-holders fixed to the terminal blocks of the dynamo.

The potential difference between any two points at any epoch is determined by means of a Kelvin quadrant electrometer and a revolving contact-maker fixed to the shaft of the dynamo. The contact-maker consists of a disc of gun-metal which carries two rings, one of gun-metal insulated from the disc, the other of ebonite. Into the latter is inserted a strip of metal $\frac{5}{64}$ inch wide, which is in permanent contact with the gun-metal ring. Two insulated brushes are attached to a moveable brush-holder, so that one presses on each ring. The circuit between the two brushes is completed once in each revolution, and the position of the contact can be read off by a pointer attached to the holder on a circle $13\frac{1}{2}$ inches diameter, divided into 360 degrees. The two points between which it is desired to measure the potential difference are connected through the contact-maker to a condenser and the quadrant electrometer, as shown in fig. 2, in which

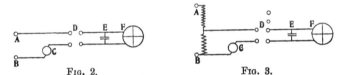

Fig. 2.   Fig. 3.

A and B are the points, C the revolving contact-maker, D the reversing switch of the electrometer, E the condenser, and F the quadrant electrometer. By plotting as ordinates the volts measured at any epoch, and as abscissæ the position of the contact-maker as representing time, the curve of potential is obtained. The electrometer is standardised by means of a Clark cell, so that the deflections on the scale can be reduced to volts: when the potential difference between A and B was too great for the electrometer, it was reduced in any desired ratio by two considerable non-inductive resistances introduced between A and B, as shown in fig. 3.

The characteristic curve of the alternator is given in fig. 4, and shows the relation between the total induction I, between one pole piece and the opposing one, in terms of the line integral of magnetising force due to the two windings in series on the two respective magnet cores; the scale of ampères in the magnet winding is also given horizontally.

162 ON ALTERNATE CURRENT DYNAMO-ELECTRIC MACHINES.

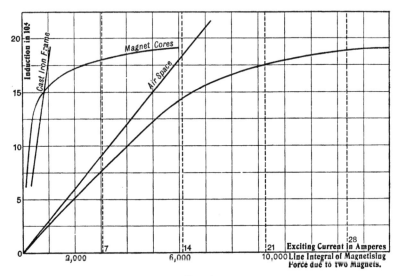

FIG. 4.

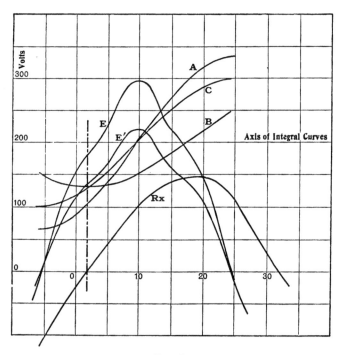

FIG. 5.

In fig. 5 the speed of the alternator experimented upon was 923 revolutions per minute, or a frequency of 92·3 complete periods per second. For the purpose of obtaining a marked effect from the current in the armature a large current was taken out of the armature, and the current in the magnet winding was only 8 ampères. A Kelvin multicellular voltmeter placed across the terminals of the machine read 190 volts on open circuit, and 81 volts when loaded, and a Kelvin ampère balance in the external non-inductive circuit read 40 ampères. The armature bobbins were coupled 6 in series 2 parallel between the brushes, the total resistance of the circuit (R) was 2·52 ohms, the armature resistance alone being ·55 ohm.

Curve E is the electromotive force curve of the machine when there is no current in the armature. Rx is the curve of electromotive force deduced from the potential difference taken between the terminals of the alternator when supplying current through non-inductive resistances. The curves E and Rx have been integrated, the corresponding integral curves being A and B respectively.

That the ordinary theory does not fully account for the facts is easily shown. We have $Rx = E - (Lx)$. Integrate both sides from any fixed epoch 0 to any time t and we have

$$\int_0^t Rx\,dt = \int_0^t E\,dt - \Big[Lx\Big]_0^t.$$

Each term of this equation consists of a constant part and of a periodic part. The constant parts must be equal and also the periodic parts. We have to deal only with the periodic parts. The curves A and B, fig. 5, represent the periodic parts of each of the first two terms; the difference of ordinates of these curves should at all times be equal to Lx. In particular, and this is the only point of moment, as we do not know how L may vary, when Lx vanishes $\int Rx\,dt = \int E\,dt$.

If the effect of the current in the armature on the induction through the armature could be represented by a term which vanishes with the current in the armature, it is obvious that the curves A and B would cross at the epoch when $x = 0$. They do not. The difference of inductions as given by these curves at this epoch is 25 per cent. of the induction which would *then* traverse the armature coil if the machine were running on open circuit, that is to say, 25 per cent. of the ordinate at this epoch given by the A curve. If it is assumed that the *average* effect of the armature current upon the induction in the magnets is such as to lower this induction by 25 per cent., the ordinates of the E curve would be decreased in like proportion, giving the curve E' of which C is the integral. The difference at any epoch between the curve C and the Rx curve is to be fully accounted for by the term (Lx).

We may put it in this way. In this machine the armature current at the times when it has a value affects the field at the instant when the armature current is zero. The effect is produced by variations induced in the current in the field magnet-winding and in the solid iron of the magnets by the varying current in, and the varying position of, the armature. That these induced currents must exist is obvious, and it is easy enough to measure them in the copper coils. They have the effect of causing the armature reaction to produce an average effect upon the magnetism of the fields by partially annulling its periodic effect. If the current in the armature did not lag behind the electromotive force of the magnets E, there would be little or no diminution of the average magnetism of the magnets. We may correctly say that this diminution of the magnetism of the magnets is due to the self-induction of the armature causing a lag of current. The effect arises from the self-induction of the armature modified by currents induced in the magnets.

The effect on the magnets of any current in the armature is greatest when the armature bobbins are opposite the pole-faces, it is *nil* or small when half-way between the pole-faces. We may therefore represent its effect at any instant approximately as proportional to the expression

$$\frac{2\pi L/T \sin 2\pi t/T + R \cos 2\pi t/T}{(2\pi L/T)^2 + R^2} \sin 2\pi t/T,$$

or

$$\frac{\pi L/T}{(2\pi L/T)^2 + R^2} - \frac{\pi L/T \cos 4\pi t/T - R \sin 4\pi t/T}{(2\pi L/T) + R^2}.$$

The constant term causes the fall in average magnetism, the periodic term causes currents in the magnets, and its effect on the magnetism is partly annulled thereby. The effect will vary as the square of the current if this is small. The effect of self-induction in diminishing the apparent electromotive force of the machine varies as the square of the current, so that we may expect the two effects, that due to the reaction of the armature on itself and that brought about by the armature inducing currents in the magnets, to vary together.

The lag of phase is less than we should expect from the diminution of electromotive force, or the electromotive force suffers greater diminution than we should expect from the angle of lag of phase.

We have tried a number of experiments for the purpose of tracing the variations of current in the magnets, and also with the intention of increasing or diminishing the effect we have observed. It is easy enough to trace the variations in the current round the magnets by measuring at points during the period the potential difference between the ends of a non-inductive resistance in series with the magnets. These variations are shown in figs. 6 and 7, in which the armature bobbins were coupled, 4 series, 3 parallel, and 12 series, respectively. We see, as we should expect, that the variations have a periodic time one-half the periodic time of the machine. But the current round the magnet could be made constant by exciting the two machines with the same current, loading each to the same degree, and placing their armatures one-fourth part of a period apart in phase.

The machines were run under conditions set forth in Table I., and the curves are given in figs. 5, 6, 7, 8. An electromotive force, curve E, was observed with no current in the armature, and a curve of potential difference was taken between the brushes with the alternator loaded on a non-inductive resistance. Rx is this curve *with* variations in the exciting current, and Rx' *without* variations, in each case corrected for the resistance of the armature. Figs. 6 and 7 show the curves of actual electromotive force, Rx and Rx', when the current in the magnet-winding is allowed to vary and when its variations are stopped. It will be seen that they do not differ a great deal. What currents are stopped in the magnet-winding no doubt turn up in the substance of the cores themselves and have an effect not differing greatly. Curve E represents electromotive forces observed. Rx represents the potential difference taken between the brushes and corrected for

166 ON ALTERNATE CURRENT DYNAMO-ELECTRIC MACHINES.

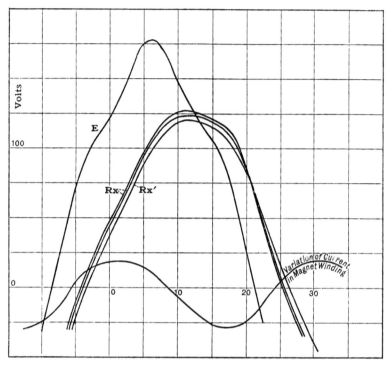

Fig. 6.

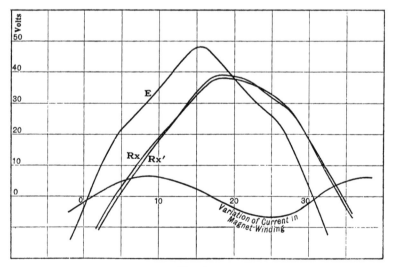

Fig. 7.

the resistance of the armature, with the alternator working on a non-inductive resistance *with* variations in the exciting current, Rx' *without* such variations in each case. The induced currents are in either case adequate to nearly stop the variation of induction. Fig. 8 shows the same thing.

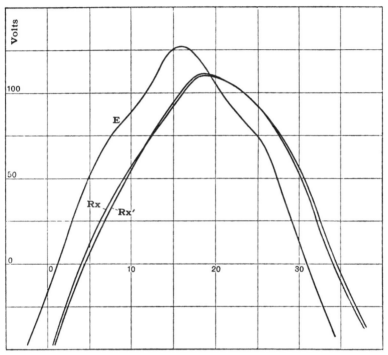

FIG. 8.

An exploring coil was wound and placed on one of the magnet limbs and the electromotive force in it was observed in terms of the time for the various positions of the exploring coil, marked 1, 2, 3, 4, 5, in fig. 1. Both the amplitude and the epoch varied with the position of the coil, but, in all cases, the periodic time was half the periodic time of the machine. It does not seem worth while to publish the curves connecting electromotive force and time.

Lastly, we tried to exaggerate the effects; for this purpose we introduced plates of copper, $\frac{1}{8}$ inch thick, in the form of two flat rings between the pole faces and the armature. Curves 9, 10, 11, 12, give the results for two different currents round the magnets

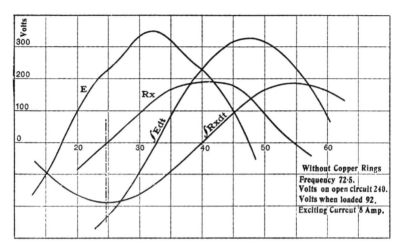

Fig. 9.*

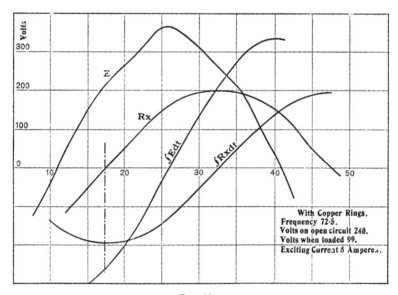

Fig. 10.

* In the case of figs. 9, 10, 11 and 12, the armature bobbins were coupled all in series, and the magnet-cores were separated in order to admit the copper plates, the air-space being thereby increased from $1\frac{1}{4}$ inches to $1\frac{3}{8}$ inches. In figs. 9 and 10 the resistance R was 7·67 ohms, and in figs. 11 and 12 it was 2·51 ohms. [Ed.]

with the copper plates in and the copper plates absent. A comparison shows that the copper plates do not make a great deal of

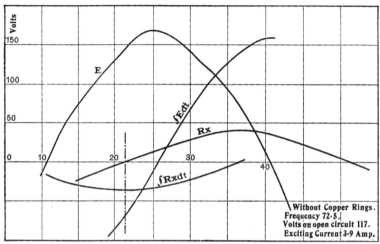

Fig. 11.

difference. The principal effect is to diminish the current induced in a coil on the magnet placed at position 5, fig. 1, close behind the copper plate.

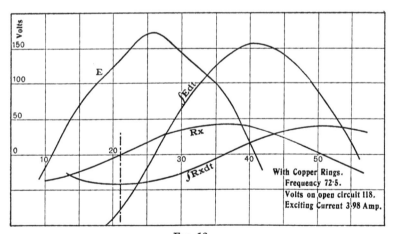

Fig. 12.

It may be inferred that in this machine there is conductivity enough in the magnet cores to have in large measure the effect indicated, and that the effect cannot be greatly diminished by compelling the magnetising current to be constant in the magnet coils, nor can it be greatly increased by exaggerating the currents induced about the magnets by intentionally introducing additional conductivities around them. The effect of each is merely to alter the place where the currents occur.

Recently machines have been built, with finely-divided pole pieces to the magnets by Messrs Mather and Platt and by the British Thomson-Houston Company.

It was obviously desirable to obtain a verification with a machine of totally different construction. For this purpose we had available the first model made of the alternating machines manufactured by Messrs Mather and Platt. It has an iron core in the armature which projects and extends beyond the armature coils. It was treated in exactly the same way as the Siemens' machines but with a fairly full load. The results are shown in fig. 13, from which it will be observed that the total induction actually observed when the machine is loaded is about 11 per cent. below the induction inferred from the electromotive force on open circuit.

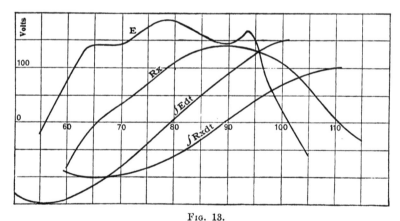

Fig. 13.

II.

The following experiments were primarily made for the purpose of determining the efficiency of the machine, but they will be seen in Section III. to have an important bearing upon the principal subject of this paper.

For the purpose of finding the efficiency of the alternators, when running as generator and motor*, the two armatures were rigidly mechanically coupled together, the leading machine being generator, and were connected in series with a non-inductive resistance r, and a Kelvin ampère balance C, as shown in fig. 14. The

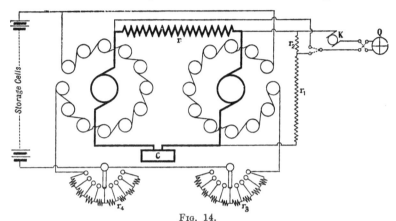

Fig. 14.

potential difference of the generator was measured at different epochs by means of the Kelvin quadrant electrometer Q, and the contact-maker K, the potential applied to the electrometer being reduced by the non-inductive resistances r_1, r_2. For corresponding epochs a curve of potential was taken across r, this gives the current passing between the machines and also the difference of potential difference between motor and generator.

The power difference, or loss in the combination, was supplied by a shunt-motor through shafting and belts, and was determined by observing the watts supplied to the motor when driving the shafting, the alternator belt being removed, and then observing the watts taken to drive the alternator when loaded—the speed being the same in each case. The difference gives the power absorbed by the combination.

It was found that the watts required to drive the shafting alone were 1681; the watts required to drive the shafting and alternators when excited, but not loaded, were 2479, the difference being in part due to currents induced in the metal frames of the armature.

* This is the same method of test which has been applied to direct current machines (see *Phil. Trans.*, R.S., 1886, p. 331).

Tables II. and III. give for about half and full load (with regard to current only) the data for getting at the watts given out by generator and received by motor, and have been obtained from the potential and current curves. The phase difference between the armatures was $\frac{1}{20}$th and $\frac{1}{10}$th period respectively.

Table IV. shows how the efficiencies of generator, motor, and combination have been obtained; and also gives the allocation of losses in the system.

The frequency was about 70 periods per second, and, since the machines are built for 100 periods per second, the figures must be taken only as illustrative of the method of test.

The alternators being connected, as shown in fig. 14, we were able to vary the exciting currents of the two machines by means of the adjustable resistances r_3, r_4. The armatures were coupled $\frac{1}{10}$th of a period apart in phase, and the following experiments were made.

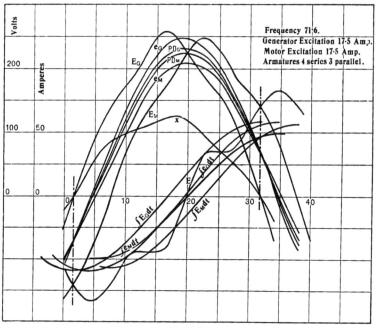

Fig. 15.

1. The alternators were equally excited with a current of 17·5 ampères and run at a speed of 716 revolutions per minute, corresponding with a frequency of 71·6 periods per second, and the following curves obtained (see fig. 15).

ON ALTERNATE CURRENT DYNAMO-ELECTRIC MACHINES. 173

E_G, E_M are the E.M.F.'s of generator and motor when running on open circuit.

PD_G, PD_M are the potential differences of generator and motor respectively when a current of 42·2 ampères ($\sqrt{\text{mean}^2}$) was passing through the armatures.

e_G, e_M are the E.M.F.'s of the respective machines when loaded, that is to say, they are the curves PD_G, PD_M corrected for current into armature resistance.

E is the E.M.F. of the combination when not loaded, that is, it is the difference of the curves E_G, E_M.

x is the curve of current passing between the machines, and is proportional to the E.M.F. (e) of the combination when loaded, the connecting leads being non-inductive. This electromotive force is the difference of the curves e_G, e_M.

2. In fig. 16* the frequency is 71, the motor is excited with 18·6 ampères and the generator with 7·2 ampères. In this case

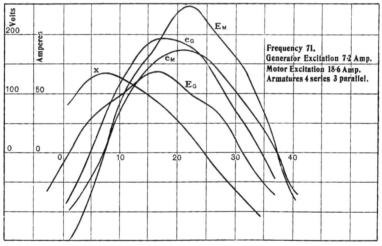

FIG. 16.

the motor is working at higher E.M.F. on open circuit than the generator. Curves corresponding to those in the Experiment 1 were obtained and are marked in a similar manner.

* *Vide* Hopkinson, Institute of Civil Engineers' Lecture delivered 1883; Institute of Civil Engineers, November, 1884; or *Original Papers on Dynamo Machinery*, pp. 58 and 148.

The potential curves in fig. 16 have been integrated, and the integral curves so obtained are given in fig. 17. These give therefore the inductions in terms of the time.

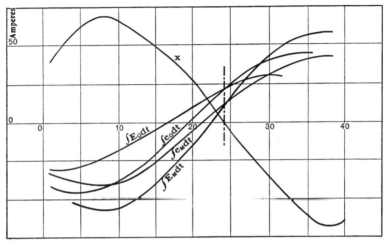

Fig. 17.

3. In fig. 18 the frequency is 70, the generator is excited with 18·6 and the motor with 11·5 ampères. The generator is working with a higher E.M.F. on open circuit than the motor. The potential curves have been integrated, and the integral curves are given in fig. 19.

It was observed that when the exciting current of the motor was decreased, that of the generator being kept fairly constant (the two machines being equally excited with about 18 ampères each to begin with), the current between the machines gradually decreased until a critical point was reached, when a further diminution of the motor exciting current had the effect of increasing the current between the machines*. It was also observed that the watts given out by the generator did not vary in the proportion of the currents between the machines.

Table V. gives the watts given out by generator for three values of its exciting currents 18·2, 18·6, and 19 ampères, the corresponding exciting currents for the motor being 18, 11·5, and 8·4 ampères. The current between the machines is a minimum

* This effect has been independently observed by Mr Mordey and Mr Kapp (see *Journal of Institute of Electrical Engineers*, vol. XXII, pp. 128, 173).

ON ALTERNATE CURRENT DYNAMO-ELECTRIC MACHINES. 175

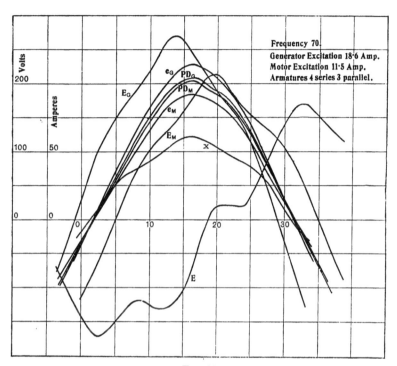

Fig. 18.

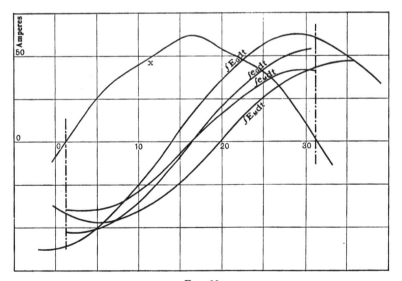

Fig. 19.

for exciting currents of 18·6 and 11·5 in the two machines, and the curves in fig. 17 have been taken under these conditions.

This is a point of practical importance in transmission of power by alternate currents, since the size of the conductor between motor and generator is mainly determined by the current transmitted. The cause is readily explained from the curves.

Starting with the conditions as in Experiment 2, where the motor is more highly excited than the generator, we see that the current (x) is accelerated in phase with regard to the potential difference of the machines. On increasing the exciting current of the generator until the machines are equally excited as in Experiment 1, the current (x) is still accelerated with regard to the potential difference, but not to such an extent. On diminishing the motor exciting current until the machines are excited as in Experiment 3, the current (x) is in phase with the potential difference. For a given power transmitted this will be the point of maximum efficiency with regard to the intermediate conductors. Any further diminution of the motor exciting current has the effect of retarding the current (x) with regard to the potential difference, and consequently for the same watts transmitted and the same potential difference the current must be increased. The case in which the conductors between motor and generator have considerable induction or capacity has not been worked out.

The losses in the system can be supplied electrically (instead of by belt as in these experiments) as in the case of direct current machines*.

III.

The results of the last section are valuable in relation to the effects of induced currents in the magnets, the subject of Section I. With a machine working as a simple generator the current lags behind the electromotive force on open circuit by any amount from 0° to 90°. But when a generator and motor are run rigidly coupled together, the current may lead the generator electromotive force, or may lag and the motor may lag by any amount from 90° to 270°. Regarding the relative phases of electromotive force of machines and of current, the machine is a generator when

---

* See *Engineering*, 24th March, 1893.

the current is from 0° to 90° behind the machine; it is a motor from 90° to 270°, and again a generator from 270° to 360°. We have already stated that we should naturally expect that the induced currents in the magnets would have little or no effect when E.M.F. and current were in the same phase, and that they would have a maximum effect when the two were 90° apart, or at quarter centres. We should expect further that, as a generator can be made into a motor by reversing the current in the armature, wherever local currents diminish E.M.F. of a generator, they would increase E.M.F. of a motor. That is, we should expect that local currents would diminish E.M.F. from lead 0° to 180°, and increase E.M.F. from 180° to 360°. As a fact, we find this to be partially verified; it seems that local currents diminish E.M.F. from a negative angle of comparatively small amount, perhaps 30°, to considerably more than 90°, and that they increase E.M.F. from 180° to over 270°.

Referring to the curves in fig. 14, x and E_G are in phase, and $\int e_G dt$ would need increasing 3 per cent. to meet $\int E_G dt$, when x vanishes. E_M lags 216°, and $\int e_M dt$ needs diminishing, that is the currents have increased the E.M.F. In fig. 15 a very small change in the observations would change the character of the results.

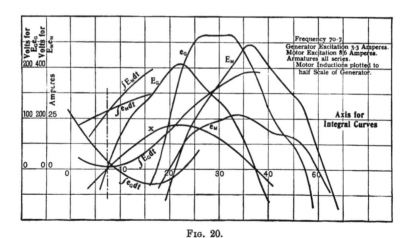

FIG. 20.

We have taken another set of curves shown in figs. 20 and 21. In fig. 20, we have a very small current 3·3 ampères in the generator magnets, and the current is in phase with the generator, the motor being 264° behind the current. The generator is about 12 per cent. low owing to local currents, and the motor is 25 per cent. of its actual value high. To obtain a better standard we excited a machine with 3·3 ampères and passed the same current as before through the armature and an ordinary non-inductive

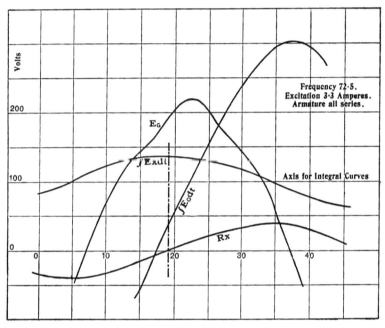

Fig. 21.

resistance, and found the current lagged 72°, and the E.M.F. of the machine was diminished 50 per cent. instead of 12 per cent. The results are given in fig. 21.

In considering the applications of these results, it must be remembered that the machines have been worked far outside the limits of practice for the purpose of accentuating the effect. If we confined ourselves to these limits we should still find these effects, but smaller in amount.

TABLE I.

Fig.	Frequency	$*\dfrac{b}{a}$		Ampères ($\sqrt{\text{mean}^2}$) per armature bobbin.		Exciting current in winding on magnets. Ampères		Remarks
		With variations in current in magnet winding	Without variations in current in magnet winding	With variations in current in magnet winding	Without variations in current in magnet winding	Ampères when normal	Maximum variation both sides from normal	
							per cent.	
5	92·3	·255	...	20	...	8	...	
	97·7	·149	...	15	...	7	...	
⎰6	95	·13	·147	13·6	13·6	7	22	External resistance altered so as to give same current
⎱6	95	·104	·147	14·1	13·6	7	22	External resistance same in each experiment
7	9·2	·093	·085	8·9	9·1	6	21	External resistance same in each experiment
8	11·7	·05	·035	17·6	17·6	16	20	External resistance same in each experiment

* $a = \tfrac{1}{2}\int_0^{\frac{T}{2}} E\,dt - \int_{E=0}^{x=0} E\,dt$ = ordinate of curve A when $x=0$.

$b = a - \tfrac{1}{2}\int_0^{\frac{T}{2}} Rx\,dt$ = difference of ordinates of curves A and B when $x=0$, which should be zero on the usual theory.

TABLE II.—*Efficiency Test of Siemens' W. 12 Alternators. Frequency 70·3 periods per second. Half Load.*

Angle in 60ths of a period	Potential at terminals of generator		Current between machines			Potential at terminals of motor		Watts given out by generator	Watts given to motor
	Electrometer deflection	Volts	Electrometer deflection	Ampères	Sq. of ampères $\sqrt{\text{mean}^2} = 21\cdot57$ Thomson balance } 21·63	Difference of potential difference	Volts		
0	−159	−224	+ 96	+23·1	533·6	+2·57	−221·4	−5174	−5114
2	−138	−194·5	+ 76	+18·3	334·9	+2·03	−192·5	−3559	−3523
4	−119	−167·7	+ 62	+14·9	222·0	+1·66	−166·0	−2499	−2473
6	− 99	−139·5	+ 44	+10·6	112·3	+1·18	−138·3	−1479	−1568
8	− 71	−100·0	+ 17	+ 4·1	16·8	+0·45	− 99·5	− 410	− 408
10	− 34	− 47·9	− 18	− 4·3	18·5	−0·48	− 48·4	+ 206	+ 208
2	+ 6	+ 8·4	− 50	−12·0	144·0	−1·34	+ 7·1	− 101	− 85
4	+ 46	+ 64·8	− 79	−19·0	361·0	−2·11	+ 62·7	−1231	−1191
6	+ 79	+111·3	− 95	−22·9	524·4	−2·54	+108·8	−2548	−2491
8	+104	+146·5	−101	−24·3	590·5	−2·7	+143·8	−3560	−3494
20	+124	+174·7	−107	−25·8	665·7	−2·86	+171·8	−4507	−4432
2	+145	+204·3	−116	−27·9	778·4	−3·1	+201·2	−5700	−5614
4	+166	+233·9	−127	−30·6	936·4	−3·4	+230·5	−7158	−7053
6	+179	+252·2	−129	−31·1	967·2	−3·45	+248·7	−7842	−7734
8	+174	+245·2	−116	−27·9	778·4	−3·1	+242·1	−6840	−6754
					6984·1			52402	51726
					465·6			3493	3448

ON ALTERNATE CURRENT DYNAMO-ELECTRIC MACHINES. 181

TABLE III.—*Efficiency Test of Siemens' W. 12 Alternators. Frequency 69·2 periods per second. Full Load.*

Angle in 60ths of a period	Potential at terminals of generator		Current between machines			Potential at terminals of motor		Watts given out by generator	Watts given to motor
	Electrometer deflection	Volts	Electrometer deflection	Ampères	Sq. of ampères $\sqrt{\text{mean}^2} = 42.8$ Thomson balance $\}$ 42·0	Difference of potential difference	Volts		
0	− 64	− 90·2	+ 25	+ 6·02	36·24	+0·67	− 89·53	− 543	− 539
2	− 35	− 49·3	− 34	− 8·2	67·24	−0·91	− 50·21	+ 404	+ 412
4	− 1	− 1·4	− 100	− 24·1	580·7	−2·67	− 4·07	+ 34	+ 98
6	+ 32	+ 45·1	− 152·4	− 36·7	1347·0	−4·07	+ 41·03	− 1655	− 1506
8	+ 63	+ 88·8	− 189·5	− 45·6	2079·4	−5·06	+ 83·74	− 4049	− 3819
10	+ 94	+ 132·5	− 211	− 50·8	2580·8	−5·64	+ 126·86	− 6731	− 6445
2	+ 121	+ 170·5	− 222·4	− 53·6	2873·0	−5·95	+ 164·55	− 9140	− 8819
4	+ 142	+ 200·1	− 236	− 56·8	3226·0	−6·31	+ 193·79	− 11367	− 11009
6	+ 155·3	+ 218·8	− 251·5	− 60·6	3672·0	−6·73	+ 212·07	− 13260	− 12850
8	+ 161·2	+ 227·1	− 251·5	− 60·6	3672·0	−6·73	+ 220·37	− 13763	− 13354
20	+ 160·2	+ 225·7	− 228·2	− 55	3025·0	−6·11	+ 219·59	− 12413	− 12080
2	+ 152·4	+ 214·7	− 188·5	− 45·4	2061·0	−5·04	+ 209·66	− 9746	− 9518
4	+ 138	+ 194·4	− 147·4	− 35·5	1260·3	−3·94	+ 190·46	− 6902	− 6761
6	+ 118	+ 166·3	− 113	− 27·2	739·8	−3·02	+ 163·28	− 4523	− 4441
8	+ 92	+ 129·6	− 75	− 18·1	327·6	−2·01	+ 127·59	− 2346	− 2309
					27548·08			96000	92940
					1836·54			6400	6196

Table IV.

No.	Description of Magnitude	Half load	Full load
	Frequency in complete periods per second	70·3	69·2
	Phase difference between armatures in fractions of a complete period	$\frac{1}{20}$	$\frac{1}{10}$
1	Watts given out by generator (see Tables II. and III.)	3493	6400
2	Watts given to motor (see Tables II. and III.) . . .	3448	6196
3	Watts dissipated in generator armature $=(\sqrt{\text{mean}^2}\, C.)^2$ ·275 ohm	128·5	481·2
4	Watts dissipated in motor armature $=(\sqrt{\text{mean}^2}\, C.)^2$ ·275 ohm	128·5	481·2
5	Watts dissipated in generator magnet winding . . .	537	537
6	Watts dissipated in motor magnet winding	537	537
7	Watts dissipated in connection between machines . .	45	204
8	Watts absorbed by combination through belt	1848	2941
9	Total electrical power developed in generator = No. 1 + No. 3	3621	6881
10	Half watts absorbed by system *minus* half known watts $=\frac{1}{2}\{\text{No. 8} - (\text{No. 3} + \text{No. 4} + \text{No. 7})\}$	746	889
11	Total power given to generator = No. 5 + No. 9 + No. 10	4904	8307
12	Percentage efficiency of generator $=\dfrac{\text{No. 1}}{\text{No. 11}} \times 100$ . .	*71·2*	*77·0*
13	Percentage loss in generator armature	2·61	5·79
14	Percentage loss in generator magnet winding . . .	10·9	6·46
15	Percentage sum of all other losses in generator . . .	15·29	9·75
16	Percentage efficiency of motor $=100\left(\dfrac{\text{No. 9} + \text{No. 10} - \text{No. 8}}{\text{No. 2} + \text{No. 6}}\right)$	*63·2*	*71·7*
17	Percentage loss in motor armature	3·22	7·14
18	Percentage loss in motor magnet winding	13·5	7·98
19	Percentage sum of all other losses in motor	20·1	13·18
20	Percentage efficiency of combination = (No. 12 × No. 16) $\frac{1}{100}$	*45·0*	*55·2*

ON ALTERNATE CURRENT DYNAMO-ELECTRIC MACHINES. 183

TABLE V.

Angle in 60ths of a period	Current in genr. mags. 18·2, motor 18, √mean² volts on motor 155. Frequency 70				Current in genr. mags. 18·6, motor 11·5, √mean² volts on motor 148. Frequency 70				Current in genr. mags. 19, motor 8·4, √mean² volts on motor 132·5. Frequency 70			
	Volts √mean² = 158	Current Electrometer deflection	Current Ampères √mean² = 43·3	Watts given out by generator	Volts √mean² = 146	Current Electrometer deflection	Current Ampères √mean² = 40·5	Watts given out by generator	Volts √mean² = 135·5	Current Electrometer deflection	Current Ampères √mean² = 43·6	Watts given out by generator
0	− 53·3	44	− 9·9	+ 528	− 33·8	42	+ 9·45	− 319	− 23·8	124	+ 27·9	− 664
3	+ 13·6	155·3	− 35·0	− 476	+ 23·3	51	− 11·48	− 267	+ 34·4	4	+ 0·9	+ 310
6	+ 75·9	219·4	− 49·4	− 3,750	+ 80·3	138	− 31·06	− 2,494	+ 81·2	88	− 19·8	− 1,608
9	+ 146·8	249·5	− 56·2	− 8,250	+ 134·2	182·5	− 41·08	− 5,513	+ 127·5	143·5	− 32·3	− 4,118
12	+ 192·6	263·5	− 59·3	− 11,420	+ 179·6	222·3	− 50·05	− 8,988	+ 167·5	202·2	− 45·5	− 7,622
15	+ 219·7	277	− 62·4	− 13,710	+ 206·2	265·5	− 59·76	− 12,320	+ 188·4	263·5	− 59·3	− 11,170
18	+ 224·2	251	− 56·5	− 12,670	+ 205·2	267·5	− 60·22	− 12,360	+ 184·2	282·8	− 63·7	− 11,730
21	+ 203·2	188·4	− 42·4	− 8,616	+ 186·6	231·2	− 52·04	− 9,710	+ 165·4	263·5	− 59·3	− 9,808
24	+ 166·6	128	− 28·8	− 4,798	+ 150·6	194·2	− 43·72	− 6,584	+ 132·3	245·5	− 55·3	− 7,316
27	+ 108·6	53	− 11·9	− 1,292	+ 96·2	141·5	− 31·85	− 3,064	+ 80·4	205	− 46·1	− 3,706
				64,454				61,919				57,432

11.

AN UNNOTICED DANGER IN CERTAIN APPARATUS FOR DISTRIBUTION OF ELECTRICITY.

[From the *Philosophical Magazine*, September, 1885.]

MANY plans have been proposed, and several have been to a greater or less extent practically used, for combining the advantage of economy arising from a high potential in the conductors which convey the electric current from the place where it is generated with the advantages of a low potential at the various points where the electricity is used. A low potential is necessary where the electricity is used; partly because the lamps, whether arc or incandescent, each require a low potential, and partly because a high potential may easily become dangerous to life. Among the plans which have been tried for locally transforming a supply of high potential to a lower and safer, the most promising is by the use of secondary generators or induction coils. It has been proved that this method can be used with great economy of electric power and with convenience; under proper construction of the induction coils it may also be perfectly safe. It is, however, easy and very natural so to construct them that they shall be good in all other respects but that of safety to life—that they shall introduce an unexpected risk to those using the supply.

In a distribution of electricity by secondary generators, an alternating current is led in succession through the primary coils of a series of induction coils, one for each group or system of lamps. The lamps connect the two terminals of the secondary

AN UNNOTICED DANGER.

coil of the induction coils. It is easy to so construct the induction coils that the difference of potential between the terminals of the secondary coils may be any suitable number of volts, such as 50 or 100; while the potential of the primary circuit, as measured between the terminals of the dynamo machine, may be very great, e.g., 2,000 or 3,000 volts. If the electromagnetic action between the primary and secondary coils, on which the useful effect of the arrangement depends, were the only action, the supply would be perfectly safe to the user so long as apparatus with which he could not interfere was in proper order. But the electromagnetic action is not the only one. Theoretically speaking, every induction coil is also a condenser, and the primary coil acts electrostatically as well as electromagnetically upon the secondary coil. This electrostatic action may easily become dangerous if the secondary generator is so constructed that its electrostatic capacity, regarded as a condenser, is other than a very small quantity.

Imagine an alternate current dynamo machine, A, fig. 1, its terminals, B, C, connected by a continuous conductor, BDC,

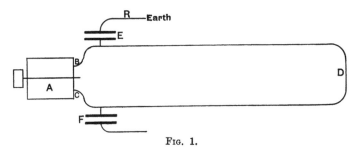

Fig. 1.

on which may be resistances, self-induction coils, secondary generators, or any other appliances: at any point is a condenser, E, one coating of which is connected to the conductor, or may indeed be part of it; the other is connected to earth through a resistance, R. Let K be the capacity of the condenser, V the potential at time t of the earth coating of the condenser, U the potential of the other coating, x the current in resistance R to the condenser from the earth, being taken as positive, and the earth potential as zero. We have

$$x = \frac{V}{R},$$

and
$$K(\dot{U} - \dot{V}) = x;$$
whence, since
$$U = A \sin 2\pi nt,$$
where A is a constant depending on the circumstances of the dynamo circuit as well as the electromotive force of the machine, and n is the reciprocal of the periodic time of the machine, we have
$$KR\dot{x} + x = 2\pi n \, KA \cos 2\pi nt,$$
$$x = \frac{2\pi n \, KA}{(KR 2\pi n)^2 + 1} \{-2\pi n \, KR \sin 2\pi nt + \cos 2\pi nt\},$$
$$\text{mean square of } x = \frac{2\pi nK}{\sqrt{(KR\, 2\pi n)^2 + 1}} \cdot \text{mean square of } A.$$

Let us now consider the actual values likely to occur in practice. Let the condenser E be a secondary generator, let the resistance R be that of some person touching some part of the secondary circuit, and also making contact to earth with some other part of the body; n may be anything from 100 to 250, say 150; K will depend on the construction of the secondary generator—it may be as high as 0·3 microfarad or even more, but there would be no difficulty even in large instruments in keeping it down to one hundredth of this or less. The mean square of A will depend on the circumstances of other parts of the circuit; it might very easily be as great, or very nearly as great, as the mean difference of potential between the terminals of the machine if the primary circuit were to earth at C. Suppose, however, that the circuit BDC is symmetrical, that E is at one end, and that another person of the same resistance as the person at E is touching the secondary circuit of the secondary generator F at the other end of the circuit. In that case, if 2,400 be difference of potential of the machine, mean square of A will be 1,200; in which case we have, taking R as 2,000 ohms,

$$\text{mean square of } x = \frac{2\pi \times 150 \times 0\cdot 3 \times 10^{-6}}{\sqrt{(2\pi \times 150 \times 0\cdot 3 \times 10^{-6} \times 2{,}000)^2 + 1}} \times 1{,}200$$
$$= \text{about } 0\cdot 3 \text{ ampère.}$$

Experiments are still wanting to show what current may be considered as certain to kill a man, but it is very doubtful whether any man could stand 0·3 ampère for a sensible length of time. It is probable that if the two persons both took firm hold of the secondary conductors of E and F, both would be killed. If the person at F be replaced by an accidental dead earth on the secondary circuit of F, the person at E would experience a greater current than 0·3 ampère.

It follows from the preceding consideration that secondary generators of large electrostatic capacity are essentially dangerous, even though the insulation of the primary circuit and of the primary coils from the secondary coils is perfect. The moral is—for the constructor, Take care that the secondary generators have not a large electrostatic capacity, say not more than 0·03 microfarad, better less than $\frac{1}{100}$ microfarad; for the inspector, Test the system for safety. The test is very easy. Place a secondary generator of greatest capacity at one end of the line and connect its secondary circuit to earth through any instrument suitable for measuring alternate currents under one ampère; put the other end of the primary to earth; the reading of the current measuring instrument should not exceed such a current as it may be demonstrated a man can endure with safety.

12.

INDUCTION COILS OR TRANSFORMERS.

[From the *Proceedings of the Royal Society*, February, 1887.]

THE transformers considered are those having a continuous iron magnetic circuit of uniform section.*

Let A be area of section of the core;

m and n the number of convolutions of the primary and secondary coils, respectively;

R, r, and ρ their resistances, ρ being the resistance of the secondary external to the transformer;

x and y currents in the two coils;

a induction per square centimetre;

α the magnetic force;

l the length of the magnetic circuit;

$E = B \sin 2\pi (t/T)$, the difference of potentials between the extremities of the primary;

T being the periodic time.

We have
$$4\pi (mx + ny) = l\alpha; \quad\quad\quad\quad (1)$$
$$E = Rx - mA\dot{a}; \quad\quad\quad\quad (2)$$
$$0 = (r + \rho) y - nA\dot{a}. \quad\quad\quad\quad (3)$$

* For a discussion of transformers in which there is a considerable gap in the magnetic circuit, see Ferraris, Torino, *Accad. Sci. Mem.*, vol. XXXVII. 1885; also chapter on the "Theory of Alternating Currents," in this volume.

From (2) and (3),
$$nE = nRx - m(r+\rho)y. \quad\quad\quad\quad\quad (4)$$

Substituting from (1),
$$x\{n^2R + m^2(r+\rho)\} = n^2E + (l\alpha/4\pi)m(r+\rho); \quad\ldots(5)$$
$$y\{n^2R + m^2(r+\rho)\} = -nmE + (l\alpha/4\pi)nR; \quad\ldots(6)$$
$$A\dot{a} = -\frac{(r+\rho)mE}{n^2R + m^2(r+\rho)} + \frac{l\alpha R(r+\rho)}{4\pi\{n^2R + m^2(r+\rho)\}} \quad\ldots(7)$$

We may now advantageously make a first approximation. Neglect $l\alpha$ in comparison with $4\pi mx$, that is, assume the permeability to be very large; we have

$$A\dot{a} = -\frac{(r+\rho)mB\sin(2\pi t/T)}{n^2R + m^2(r+\rho)}; \quad\ldots(8)$$

$$Aa = \frac{(r+\rho)mB\cos(2\pi t/T)}{\{n^2R + m^2(n+\rho)\} \cdot \dfrac{2\pi}{T}}. \quad\ldots(9)$$

For practical purposes these equations are really sufficient.

We see first that the transformer transforms the potential in the ratio n/m, and adds to the external resistance of the secondary circuit ρ a resistance $(n^2R/m^2) + r$. This at once gives us the variation of potential caused by varying the number of lamps used. The phase of the secondary current is exactly opposite to that of the primary.

In designing a transformer it is particularly necessary to take note of equation (9), for the assumption is that a is limited so that $l\alpha$ may be neglected. The greatest value of a is $B/\{(2\pi/T)mA\}$, and this must not exceed a chosen value. We observe that B varies as the number of reversals of the primary current per unit of time.

But this first approximation, though enough for practical work, gives no account of what happens when transformers are worked so that the iron is nearly saturated, or how energy is wasted in the iron core by the continual reversal of its magnetism. The amount of such waste is easily estimated from Ewing's results when the extreme value of a is known, but it is more instructive to proceed to a second approximation, and see how the magnetic properties of the iron affect the value and phase of x and y. We shall, as a second approximation, substitute in equations (5), (6), (7) values of α deduced from the value of a furnished by the first approximation in equation (9).

In the accompanying diagram, fig. 1, Ox represents α, Oy represents a, and Oz the time t.

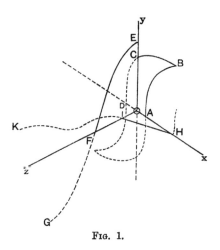

FIG. 1.

The curves $ABCD$ represent the relations of a and α. EFG the induction a as a function of the time, and HIK the deduced relation between α and t. We may substitute the values of α obtained from this curve in equations (5) and (6), and so obtain the values of x and y to a higher degree of approximation. If the values of α were expressed by Fourier's theorem in terms of the time, we should find that the action of the iron core introduced into the expression for x and y, in addition to a term in cos $(2\pi t/T)$ which would occur if a and α were proportional, terms in sin $(2\pi t/T)$ and terms in sines and cosines of multiples of $2\pi t/T$. It is through the term in sin $(2\pi t/T)$ that the loss of energy by hysteresis comes in.

A particular case, in which to stay at a first approximation would be very misleading, is worthy of note. Let an attempt be made to ascertain the highest possible values of a by using upon a transformer a very large primary current and measuring the consequent mean square of potential in the secondary circuit by means of an electrometer, by the heating of a conductor, or other such device. The value of a will be related to the time somewhat as indicated by *ABCDEFG* in fig. 2; for simplicity assume it to be as in fig. 3; the resulting relations of potential in the secondary and the time will be indicated by the dotted

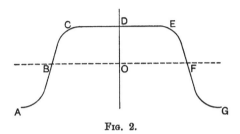

FIG. 2.

line *HIJKOLMNPQ*. The mean square observed will be proportional to $ML \cdot \sqrt{LP}$; but $ML \cdot LP$ is proportional to EL, hence the potential observed will vary inversely as $\sqrt{LP}$, even though the maximum induction remain constant. If, then, the maximum

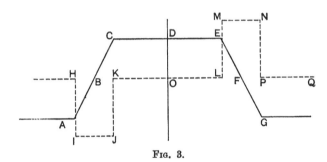

FIG. 3.

induction be deduced on the assumption that the induction is a simple harmonic function of the time, results may readily be obtained vastly in excess of the truth.

13.

REPORT TO THE WESTINGHOUSE COMPANY OF THE TEST OF TWO 6,500-WATT WESTINGHOUSE TRANSFORMERS. *May* 31, 1892.

BEFORE giving any of the results of the tests I have made with your transformers, it will be well to explain the methods of experiment adopted. The instantaneous value at any epoch in the period of the difference of potential between any two points of a circuit in which the potential difference is varied periodically is made effective on the measuring instrument by means of a rotating contact maker attached to the shaft of the alternate current generator. This contact maker was constructed for the King's College laboratory by Messrs Siemens Brothers. It makes contact once in each revolution for a period of about three-quarters of a degree, and breaks it for the rest of the revolution. It is entirely insulated, and so can be connected to any part of the circuit. The position of the contact can be varied, and the variation be read off on a graduated circle of $13\frac{1}{2}$ inches diameter divided into degrees, and by estimation the variation can be read to one-tenth of a degree. The two points between which it is desired to measure a potential difference are connected through the contact maker to a condenser and a quadrant electrometer, as shown in fig. 1, in which A and B are the points, the potential difference of which at a stated epoch is to be measured, C the revolving contact maker, D the reversing switch of the electrometer, E the condenser, of which the capacity can be varied, F the quadrant electrometer. It is evident that the quadrant electrometer will give a reading proportional to the potential difference of A and B, when C makes contact. If there were no leakage, it would at once give this potential. It is to obviate the effect of leakage that the condenser is introduced, and the amount of the effect was determined by varying the

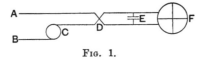

Fig. 1.

condenser thus: When the condenser had capacity 1, 0·5, and 0·2 microfarads, the readings of the electrometer for a given potential difference of an alternating current at the position in the period of maximum electromotive force were 138, 136, and 132, respectively. The rate of loss of potential will be proportional to the reciprocal of the capacity, whence we infer that the true reading, if insulation were perfect, would be $139\frac{1}{2}$, and hence the readings are always corrected by adding one per cent. When the potential difference was too great for the electrometer it was reduced in any desired ratio by two considerable resistances introduced between the points to be measured in the usual way (fig. 2). The potential difference may, of course, be measured

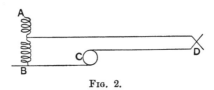

Fig. 2.

in other ways. An ordinary voltmeter may be placed between A and B, in which case it must be standardized with the contact breaker in circuit; and it will depend for its constant on the duration of the contact, which may vary. Further, it gives, not the difference of potential at any definite epoch, but the mean difference for the whole time of the contact. The condenser may be used and its potential be measured by discharge through a galvanometer: this is open to the objection that if there be any leakage, the result will depend on the time at which the contact is broken by the condenser key in relation to the time at which it was made by the revolving contact maker. Lastly, a Clark cell may be used, by a method which Major Cardew pointed out to me (fig. 3), the resistance being adjusted till there is no deflection. This is open to the same objection as the first,

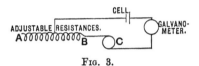

Fig. 3.

namely, that it gives the mean of the potential differences which occur during the contact. By making use of the first-mentioned method we have the means of measuring accurately any potential difference at any epoch of the period, and of knowing the epoch.

For these experiments two transformers intended to be identical were available, each transforming between 2,400 and 100 volts. It was most convenient on account of the resistance available to couple these transformers up from 100 to 2,400 in the first or No. 1 transformer, then down from 2,400 to 100 in the second or No. 2 transformer, and to take up the energy from the second in a non-inductive resistance. The arrangement is shown in fig. 4.

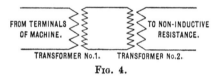

Fig. 4.

The obvious way in returning the efficiency of the combination would be to measure, at various epochs of half a period, the potential differences of the terminals of the machine and the current passing in the No. 1 transformer; in like manner, at the same epochs, to measure either the potential differences or the current passing to the non-inductive resistance, thence to deduce the power supplied to the first transformer and taken from the second. This would be open to certain objections; we are comparing two nearly equal magnitudes and desire their ratio; the ratio will be affected with the full error arising from an error in the determination of either magnitude, and such errors may be material, as the observations are not simultaneous, and conditions may change between one series of observations and another.

These objections are avoided by the method adopted. The current from No. 2 is observed at certain epochs, the difference of current between No. 2 and No. 1, and the difference of potential difference of No. 2 and No. 1 at the same epochs. These give the currents and potentials of No. 1 at the same epochs as the corresponding determinations of No. 2, and the difference will only be affected with the proportion of error of those differences. For example, suppose the efficiency of the combination were 90 per cent., and the possible error of determination of power 1 per cent., our result might be anything from 88 per cent. to 92 per cent. if made in the obvious way, but if made by differences the maximum loss would be 10·1 per cent., and the possible least determination of the efficiency would be 89·8 per cent. The method is essentially similar to the method I described* and

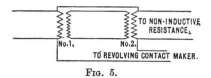

Fig. 5.

subsequently used for testing dynamos. The measurements for difference of potential differences are made as in fig. 5 and for current differences as in fig. 6, where G is a known small non-

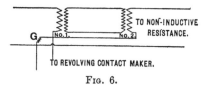

Fig. 6.

inductive resistance. The two currents will, of course, slightly disturb each other, but this is readily allowed for in the calculations.

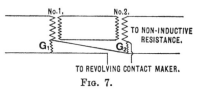

Fig. 7.

Another method would be to couple them as in fig. 7, G_1 and G_2 being equal non-inductive resistances. This arrangement

* *Phil. Trans.*, 1886, page 347.

is quite free from disturbance, but requires two resistances adjusted to exact equality. A single transformer can be tested in the same way, though in this case reliance must be placed upon resistances to reduce the current of the low potential coil, and to reduce the potential of the high potential coil in the ratio of the number of windings in the two coils.

The current was throughout generated by a Siemens alternator with 12 magnets, run at a speed between 830 and 840 revolutions per minute, which gives a frequency of 5,000 per minute, or 83 to 84 per second.

The first experiment tried* was with the two transformers coupled, but with No. 2 transformer on open circuit, or on nearly open circuit, for a high resistance for purposes of measurement was interposed between the terminals of the low resistance coil of No. 2 transformer. The actual results are given in Table I.

TABLE I.

Leads of Exploring Brush on Divided Circle	Current No. 1. Thick Coils Ampères	Potential No. 2. Thick Coils		Potential No. 1. Thick Coils		Square of Volts $\sqrt{\text{mean}^2}$ =101·9	Watts supplied to No. 1
		Volts	Square of Volts $\sqrt{\text{mean}^2}$ =101·1	P. D. See Nos. 1 and 2 Volts	Volts		
267	− 2·2	+ 25·4	645	+0·9	+ 26·3	692	− 57·9
270	− 0·3	+ 70·2	4,928	+1·2	+ 71·4	5,098	− 21·4
273	+1·1	+ 95·3	9,082	+1·1	+ 96·4	9,292	+106·0
276	+2·1	+120·4	14,496	+1·1	+121·5	14,761	+255·1
279	+2·8	+147·7	21,816	+1·1	+148·8	22,140	+416·6
282	+3·2	+147·2	21,668	+0·9	+148·1	21,935	+473·9
285	+3·4	+119·8	14,351	+0·7	+120·5	14,520	+409·7
288	+3·5	+ 97·8	9,565	+0·6	+ 98·4	9,683	+344·4
291	+3·7	+ 71·3	5,084	+0·4	+ 71·7	5,140	+260·3
294	+3·5	+ 26·0	676	+0·3	+ 25·97	674	+ 90·9
			102,311			103,935	2,282·6

* So far as I know, the first discussion of endless magnetic circuit transformers, based on the actual properties of the material, is in a note by myself (*Proc. Roy. Soc.* vol. XLII., and *The Electrician*, vol. XVIII. p. 421). Definite results were obtained by methods generally similar to those now used by Prof. Ryan (*The Electrical World*, Dec. 28, 1889). The theory of transformers is well set forth by Prof. Fleming (*The Electrician*, April 22 and 29, 1892).

and are expressed in fig. 8. Tables II., III., and IV. give the results for half power, nearly full power, and full power, and the sets of curves of figs. 9, 10 and 11 give the results of the table. In these tables the first column gives the position of the contact brush in degrees, so that 60 on this scale corresponds with a

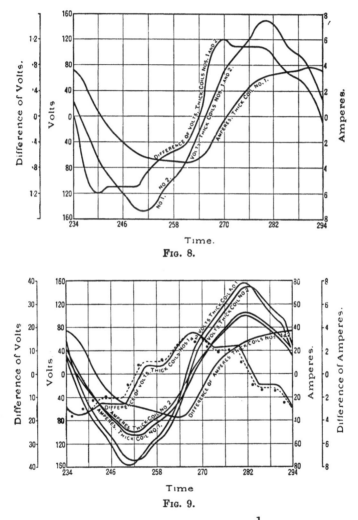

Fig. 8.

Fig. 9.

complete cycle. Three degrees are thus $\dfrac{1}{83\cdot3 \times 20}$ of a second. The second column of Table I. is the current in the thick coil of No. 1 transformer, as determined by the difference of potential at

198   TEST OF WESTINGHOUSE TRANSFORMERS.

the two ends of a non-inductive resistance in which the current passes. The third column is the potential difference of No. 2 transformer, a direct determination. The fourth column is solely for the purpose of determining the square root of the mean of the

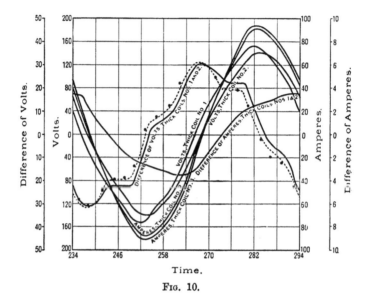

Fig. 10.

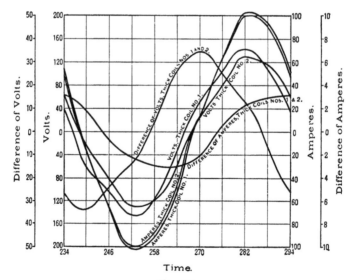

Fig. 11.

squares of the third column. The fifth column is a direct determination of the difference of potential of No. 1 and No. 2, obtained in the manner explained with reference to fig. 5.

The sixth column is the deduced potential difference of the terminals of the thick wire of No. 1 transformer, being the sum of the third and fifth columns. The seventh column, like the fourth, is merely for the purpose of determining the square root of the mean of the squares of column six, while the eighth gives the rate at which power is given out or received by the pair of transformers.

If the transformers had been exactly equal, the potentials for the two given by Table I. would have been equal, though they would have differed a little in phase owing to the lines of magnetic induction which pass through the non-magnetic space between the two coils of the transformer*. The difference shows that No. 1 transformer has a ratio of transformation slightly greater than No. 2. If we correct the potential of either No. 1 or No. 2, there still remains a difference between them, but this difference will be greatest about when the potentials are *nil*. This is due to the lost induction just referred to. In order to check the conclusion that the two transformers are not precisely equal, they were directly compared, as in fig. 12. The transformers were coupled parallel, as in fig. 12, and the difference of

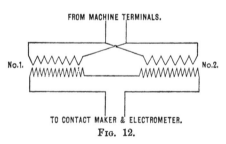

Fig. 12.

potential of the two high potential coils was measured: the value of its square root of mean square was 12·5 volts, the potential of the transformer being 2,400. This does not necessarily imply that the potentials of the two transformers differ by one half per cent.; it may be largely due to a difference of phase between the two.

* Prof. Perry has already pointed out that the effect of such an induction cannot be entirely neglected, even in endless circuit transformers

The current supplied to the No. 1 transformer is to be accounted for by the currents necessary to magnetize the two transformers, and by the local currents in their cores. To ascertain the former, the curve of magnetization of one of the transformers was determined by the ballistic galvanometer for nearly the same induction as in Table I., the changes of current supplied by a battery being made by a reversing switch, or by suddenly introducing resistance into the primary circuit, and the consequent changes of induction being measured by the galvanometer. The tardiness of change of current in the transformer due to its self induction was sufficiently reduced by using many cells and a considerable resistance. The results are shown in fig. 13

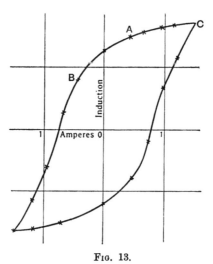

FIG. 13.

for a single transformer. In this curve the abscissæ are the currents in the thin coil of the transformer, divided by 24 to reduce it to the same effect as it would have had if it had been in the thick coil. The ordinates are the inductions as measured by the kick on the galvanometer, but reduced to a scale to make them directly comparable with the volts when the transformer is used with an alternating current. These results are not given in absolute units. The procedure to determine points on the curve was: first, pass the maximum current corresponding to the point C; next, suddenly diminish the current by inserting suitable resistance in the thin coil circuit, and observe the kick—the drop of ordinate from C to A corresponds to the kick, and the abscissa of A is the current after it has been reduced; next, reverse the current, and observe the kick—the kick corresponds to the further drop of ordinates from A to B. In this manner a series of points are determined on the curve. Fig. 14 shows the relation between

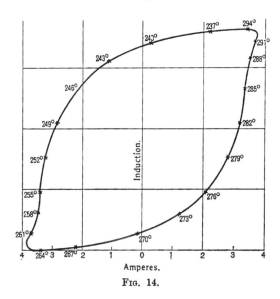

Fig. 14.

induction and magnetizing current for the pair of transformers, as deduced from the experiments with alternating currents set forth in Table I. The ordinates in this curve are the area of the curve of potentials of fig. 8, for the ordinates of this latter curve are the rates at which the induction is changing, while the abscissæ are the currents in the thick wire at corresponding times. The points marked • in fig. 15 give the remainder after deducting the magnetizing current as estimated in fig. 13 from the currents of fig. 14—that is to say, fig. 13 is corrected first for the small difference in maximum induction; then, corresponding to any induction, the current is taken from the curve, it is doubled, as there is only one transformer, and the result is deduced from the corresponding current of fig. 14. The differences are, the magnetizing current equivalent and opposite in effect to the local currents in the cores. If the local currents were equivalent to a current in a single secondary circuit, the points • of fig. 15 ought to have had the form of the full line of fig. 15, drawn through the points +, in which the abscissæ are proportional to the potential difference, and the ordinates to the induction. Returning to Table I., we find that the fall of potential difference on open circuit in the whole combination is 0·8 volt, and that the loss of power in magnetizing the cores and in local currents is 228·26 watts, that is, a loss for each transformer of 114·13 watts. The total loss of 228 watts may be divided into 126 watts accounted for by hysteresis and 102 watts due to local currents.

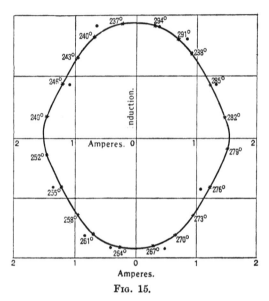

FIG. 15.

Referring now to Table III. and fig. 10, the earlier columns explain themselves, but a word is necessary about the last six columns. The watts supplied to No. 1 are simply the products at each time of the volts at its terminals and the ampères passing through, similarly to the watts given out by No. 2. We see first that the efficiency of the whole combination with this load is 93·73 per cent., and hence the efficiency of one transformer, if the losses in the two are equal, may be taken as 96·9 per cent. The fall of potential in the whole combination is 6·1 volts, but the fall with no load is 0·8 volt; hence the variation due to the load with constant potential on the thin coil of No. 1 is 5·3 volts, or, if the fall of potential in the two transformers were equal, which it is not, for a single transformer 2·65 volts. Assuming that the transformers are equal, the power lost in resistance would be expected to be the mean of mean current × the difference of potential difference, or 215·4 watts. It is, in fact, 150 watts, as given by multiplying the square of currents by resistances. But the transformers are not exactly equal, and there is the waste magnetic field, both of which will have a small effect on the distribution of loss between the two classes of loss,—viz., that by hysteresis and local currents, and that by resistance,—but none upon the gross efficiency.

The other tables, II. and IV., are arranged in exactly the same way as Table III., but the number of observations on Table IV. is insufficient to bring out all the peculiarities of the transformers.

It has already been stated that, if the loss of potential due to load in the two transformers be equal, it will amount to 2·65 per cent. The following experiment was tried to ascertain if this loss was equal: The transformers were coupled in series as before. The mean potential difference of the thick wire was measured by Thomson's multicellular, and of the thin wire by Thomson's electrostatic voltmeter. The mean of a considerable number of experiments is given in the following table, the load being the same as in Table III., and the results being corrected to the same potential of the thin wire :—

Number	Full Load		Open Circuit	
	Thomson's Electrostatic	Thomson's Multicellular	Thomson's Electrostatic	Thomson's Multicellular
1	2,380	99·8	2,380	97·0
2	2,380	94·2	2,380	96·2

This shows that of a total drop of 4·8 volts, 2·8 volts occurred in No. 1, and 2 volts in No. 2. There is no doubt of the fact that the drop is greater in No. 1 than in No. 2, which is connected with the waste field between the two coils. Of course these transformers are intended to work exactly as No. 2 is working, in which case the drop from no load to nearly full load, as shown by this experiment, is 2·0 volts. The way in which this waste field causes inequality of drop of potential in the two transformers, coupled as in my experiments, is well worthy of careful consideration. The waste field is proportional to the current in the transformers, or, better, to the mean of the two currents in ampère turns. The electromotive force due to this waste field will be proportional to the rate of change of the current. If the current were expressed by a simple harmonic curve, the electromotive force due to the waste field would also be a simple harmonic curve differing in phase by $\frac{\pi}{2}$. The curve of potentials is roughly in the same phase as the curve of current. Let a be the amplitude of potential difference of No. 2 transformer, b be the amplitude of difference between the potential difference in No. 2, and the potential difference of the thin wire divided by 24. $2b$ will be very nearly the amplitude of difference between the thick wires of Nos. 1 and 2. The ratios of potentials in No. 1 and No. 2 will then be

$$\frac{\sqrt{a^2 + 4b^2}}{\sqrt{a^2 + b^2}} \text{ and } \frac{\sqrt{a^2 + b^2}}{a}, \text{ or } 1 + \frac{3b^2}{2a^2} \text{ and } 1 + \frac{b^2}{2a^2},$$

or the drop in the first from this cause is three times as great as in the second transformer. We shall return to the waste field immediately. Putting aside harmonic curves, and returning to the facts as they are, the following table gives: first, half the

Half Difference of Potential Difference	Volts of High Potential Coil Divided by 24	Squares of Volts
15·6	− 5·0	25·0
14·8	41·3	1,706·0
11·3	74·3	5,520·0
10·8	102·6	10,530·0
11·1	131·3	17,240·0
5·5	147·1	21,640·0
− 1·1	137·0	18,770·0
− 3·1	119·3	14,230·0
− 5·9	95·1	9,040·0
−12·1	53·4	2,850·0

Square root of mean square = 100·8

difference of potential difference taken from Table III., that is, at each instant the drop of potential in No. 2; secondly, the volts of the thin wire of No. 2 reduced for number of convolutions—this is of course the mean of potential difference between 1 and 2; lastly, the squares of these volts. From this we see a mean square 100·8 showing the drop in No. 2 to be 2·6 volts out of a total drop of 6·1, and the remainder 3·5, the drop in No. 1. Diminishing these results by 0·4, the half of 0·8, the fall observed with no load, the actual losses from no load to nearly full load will be 2·2 and 3·1.

Turn now to the last column of Table III. This gives the difference of potential differences corrected for the loss of volts by resistance. It is shown dotted on fig. 10; this curve presents one or two peculiar features. It should be possible to infer the form of this curve from the curve of current. The rates at which the mean current is changing are as follows:

268½	271½	274½	277½	280½	283½	286½	289½	292½	....
30·7	24·2	19·2	18·5	13·9	1·7	−9·3	−12·7	−21·7	−28·1

—which happens to come to a scale which can be at once plotted. The points marked are the points of the curve corresponding with the above rates. The agreement of the points with the curve is remarkably close. This exhibits very completely the effect of waste magnetic field in this transformer.

For half power, as taken from Table II., the rates are as follows:

268½	271½	274½	277½	280½	283½	286½	289½	292½	295½
14·6	11·5	9·4	10·6	4·4	−4·6	−7·2	−7·5	−13·6	−17·6

—and in the same way in fig. 9 the dotted curve represents the difference of electromotive force corrected for resistance, and the points correspond with the above rates.

Fig. 16 gives the efficiencies for the combined transformers in terms of the load. This curve is the hyperbola:

$$\text{Efficiency} = 100 \cdot \frac{X - (A + BX + CX^2)}{X},$$

where $A = 228$, the loss by hysteresis;

$B = 0·005$, and mainly depends upon the waste field;

$C = 0·0000035$, and is mainly the loss by resistance;

$X = $ load in watts.

To sum up, I find that the efficiency of the transformer at full load would be 96·9 per cent.; at half load, 96 per cent.; and at quarter load, over 92 per cent. The magnetizing current of the transformer amounts to 114 watts, or 1·75 per cent. The drop of potential from no load to full load is between 2 per cent. and 2·2 per cent.

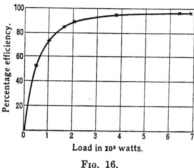

Fig. 16.

In conclusion, I wish to express my thanks to Mr Wilson, of King's College; this gentleman carried out the experiments under my direction, and made nearly all the numerical calculations and drew most of the curves for me.

TABLE II.

Lead of Exploring Brush on Divided Circle	Current No. 2. Thick Coil			Current No. 1. Thick Coil					Mean of Currents in Nos. 1 and 2. Thick Coils	Potential No. 2. Thick Coil	
	Observed Deflection	Corrected Deflection	Ampères	Current Difference. Thick Coils, Nos. 1 and 2						Volts	Square of Volts. $\sqrt{\text{mean}^2} = 104$
				Observed Deflection	Corrected Deflection	Volts	Ampères	Ampères			
267	2·5	2·5	+ ·0·6	189·5	189·0	+3·7	−2·9	−2·3	−0·8	+5·6	31
270	65·0	65·0	+16·0	77·0	77·0	+1·5	−1·2	+14·8	+15·4	+51·9	2,694
273	108·5	108·5	+26·7	28·0	28·0	−0·6	+9·4	+27·1	+26·9	+83·6	6,989
276	145·0	145·0	+35·6	99·0	99·0	−1·9	+1·5	+37·1	+36·3	+110·3	12,165
279	186·0	186·0	+45·7	152·5	152·5	−3·0	+2·4	+48·1	+46·9	+141·1	19,910
282	205·0	203·2	+49·9	186·0	186·0	−3·7	+2·9	+52·8	+51·3	+153·6	23,592
285	184·0	184·0	+45·2	203·0	201·2	−4·0	+3·1	+48·3	+46·7	+138·5	19,181
288	154·0	154·0	+37·8	220·0	218·0	−4·3	+2·4	+40·2	+39·5	+114·9	13,202
291	123·0	123·0	+30·2	238·5	235·5	−4·6	+3·6	+33·8	+32·0	+90·6	8,208
294	67·7	67·7	+16·6	237·0	234·0	−4·6	+3·6	+20·2	+18·4	+47·7	2,275
											108,247

[This table is continued on the next page.]

TEST OF WESTINGHOUSE TRANSFORMERS.

TABLE II.—*continued.*

Lead of Exploring Brush on Divided Circle	Potential Difference No. 1. Thick Coil				Mean of Potential in Nos. 1 and 2, Thick Coils	Watts supplied to No. 1	Watts given out by No. 2	Loss by Resistance and probably Waste Field		Loss by Hysteresis and Local Currents	Difference of Potential due to Waste Field
	Potential Difference Thick Coils, Nos. 1 and 2		Volts	Square of Volts. $\sqrt{mean^2} = 107.4$				Resistance	Losses unaccounted for, probably Waste Field		
	Observed Deflection	Corrected Deflection	Volts								
267	315·0	308·0	+18·2	+23·8	566	+14·7	−55	+3	0·02	−14·62	−42·3	+18·2
270	278·0	273·0	+16·1	+68·0	4,624	+60·0	+1,006	+830	8·7	+239·2	−72·0	+15·5
273	196·5	195·0	+11·5	+95·1	9,044	+89·3	+2,577	+2,232	26·6	+282·7	+35·7	+10·5
276	199·0	197·2	+11·6	+121·9	14,860	+116·1	+4,523	+3,927	48·5	+372·5	+174·1	+10·3
279	203·0	201·2	+11·9	+153·0	23,405	+147·1	+7,359	+6,448	80·9	+477·2	+353·0	+10·2
282	61·0	61·0	+3·6	+157·2	24,710	+155·4	+8,300	+7,665	96·8	+87·9	+451·6	+1·7
285	73·0	73·0	−4·3	+134·2	18,009	+136·3	+6,482	+6,260	80·3	−281·3	+422·5	−6·0
288	75·0	75·0	−4·4	+110·5	12,210	+112·7	+4,552	+4,343	57·5	−231·3	+383·0	−5·9
291	126·5	126·5	−7·5	+83·1	6,904	+86·9	+2,808	+2,736	37·7	−277·7	+312·8	−8·7
294	252·0	248·5	−14·7	+33·0	1,089	+40·3	+666	+792	12·5	−283·0	+145·1	−15·4
					115,424		38,218	35,236	449·5	371·6	2,162·4	
								92·2%		298·3		
							298·2					

TEST OF WESTINGHOUSE TRANSFORMERS.

TABLE III.

Lead of Exploring Brush on Divided Circle	Current No. 2, Thick Coil, Ampères	Current No. 1 Thick Coil — Current Difference. Secondaries. Nos. 1 and 2, Ampères	Current No. 1 Thick Coil, Ampères	Mean of Current in Nos. 1 and 2, Thick Coils	Potential No. 2 Thick Coil, Volts	Potential No. 2 Thick Coil, Square Volts, √mean² = 98·2	Potential Difference No. 1 Thick Coil — Potential Difference. Thick Coils, Nos. 1 and 2, Volts	Potential Difference No. 1 Thick Coil, Volts	Potential Difference No. 1 Thick Coil, Square Volts, √mean² + 104·4	Mean of Potential in Nos. 1 and 2, Thick Coils	Watts supplied to No. 1	Watts given out by No. 2	Loss by Resistance and probably Waste Field — Resistance	Loss by Resistance and probably Waste Field — Losses unaccounted for, probably Waste Field	Loss by Hysteresis and Local Currents	Difference of Potential due to Waste Field
267	−15·0	−3·1	−18·1	−16·5	+20·6	424	+31·2	+10·6	112	−5·0	−192	+309	10·4	−525·1	+15·5	+31·8
270	−15·0	−1·6	−13·4	−14·2	+26·5	702	+29·7	+56·2	3,305	+41·3	+753	+397	7·6	−414·1	+66·1	+29·2
273	+38·5	−0·2	+38·3	+38·4	+63·0	3,969	+22·7	+85·7	7,345	+74·3	+3,282	+2,425	55·9	+815·7	−14·9	+21·2
276	+57·1	+1·0	+58·1	+57·6	+91·8	8,427	+21·7	+113·5	12,880	+102·6	+6,595	+5,242	125·7	+1,124·3	+102·6	+19·5
279	+75·2	+1·9	+77·1	+76·1	+120·2	14,440	+22·2	+142·4	20,280	+131·8	+10,980	+9,038	219·4	+1,470·1	+249·5	+19·3
282	+88·8	+2·5	+91·3	+90·0	+141·6	20,050	+11·0	+152·6	23,285	+147·1	+13,930	+12,575	307·0	+683·0	+367·7	+7·6
285	+86·9	+2·8	+89·7	+88·3	+138·1	19,070	−2·3	+135·8	18,440	+136·9	+12,180	+12,000	295·4	+498·5	+383·3	+5·7
288	+77·5	+3·1	+80·6	+79·0	+122·4	14,980	−6·2	+116·2	13,502	+119·3	+9,365	+9,486	236·8	+726·2	+369·8	+9·2
291	+64·6	+3·4	+68·0	+66·3	+101·0	10,200	−11·8	+89·2	7,957	+95·1	+6,066	+6,524	166·5	+948·8	+323·3	−14·3
294	+42·9	+3·5	+46·4	+44·6	+65·5	4,290	−24·2	+41·3	1,705	+53·4	+1,916	+2,810	75·4	−1,154·6	+186·9	−25·9
						96,552			108,811		64,875	60,806	1,499·7	654·0	1,917·6	
												93·73 %				
						406·9			407·1							

H. I.   14

TABLE IV.

Lead of Exploring Brush on Divided Circle	Current No. 2. Thick Coil			Current No. 1. Thick Coil				Mean of Current in Nos. 1 and 2. Thick Coils	Potential No. 2. Thick Coil	
	Observed Deflection	Corrected Deflection	Ampères	Current Difference. Thick Coils. Nos. 1 and 2			Ampères		Volts	Square of Volts. $\sqrt{\text{mean}^2} = 90\cdot 9$
				Observed Deflection	Corrected Deflection	Volts	Ampères			
270	59	59	+14·3	107	107	+2·1	−1·7	+12·6	+13·4	+21·3	454
276	256	252	+61·0	57	57	−1·1	+0·9	+61·9	+61·4	+80·7	6,512
282	427	410	+99·3	147	147	−2·9	+2·3	+101·6| +100·4	+130·2	16,950
288	381	369	+89·3	188	187	−3·7	+2·9	+92·2	+90·8	+116·1	13,480
294	207	205	+49·6	211	209	−4·1	+3·2	+52·8	+56·2	+62·4	3,893
											41,289

[*This table is continued on the next page.*]

TEST OF WESTINGHOUSE TRANSFORMERS.

TABLE IV.—*Continued*.

Lead of Exploring Brush on Divided Circle	Potential No. 1. Thick Coil					Mean of Potential in Nos. 1 and 2. Thick Coils	Watts supplied to No. 1	Watts given out by No. 2	Losses by Resistance and probably Waste Field		Loss by Hysteresis and Local Currents	Difference of Potential due to Waste Field
	Potential Difference. Thick Coils. Nos. 1 and 2			Volts	Square of Volts. $\sqrt{\text{mean}^2} = 98 \cdot 4$				Resistance	Losses unaccounted for, probably Waste Field		
	Observed Deflection	Corrected Deflection	Volts									
270	355	345	+33·9	+55·2	3,047	38·3	695	304	6·8	+ 447·5	− 65·1	+33·4
276	256	252	+24·8	+105·5	11,130	93·1	6,530	4,923	142·8	+1,380·2	+ 83·8	+22·5
282	142	142	+14·0	−144·2	20,790	127·2	14,650	12,928	382·0	+1,023·5	+315·5	+10·2
288	59	59	− 5·8	+110·3	12,165	113·2	10,170	10,367	312·3	− 839·0	+328·3	− 9·2
294	266	262	−25·8	+36·6	1,340	49·5	1,933	3,095	99·3	−1,420·3	+158·4	−27·7
					48,472		33,978	31,617	943·2	591·9	820·9	
								93·05%				
						236·1					235·6	

14.

PRESIDENTIAL ADDRESS TO THE INSTITUTION OF ELECTRICAL ENGINEERS, January 9, 1890.

Magnetism.

As old as any part of electrical science is the knowledge that a needle or bar of steel which has been touched with a loadstone will point to the North. Long before the first experiments of Galvani and Volta the general properties of steel magnets had been observed—how like poles repelled each other, and unlike attracted each other; how the parts of a broken magnet were each complete magnets with a pair of poles. The general character of the earth's magnetism has long been known—that the earth behaves with regard to magnets as though it had two magnetic poles respectively near the rotative poles, and that the direction of the needle has a slow secular motion. For many years the earth's magnetism has been the subject of careful study by the most powerful minds. Gauss organised a staff of voluntary observers, and applied his unsurpassed powers of mathematical analysis to obtaining from their results all that could be learned.

The magnetism of iron ships is of so much importance in navigation that a good deal of the time of men of great power has been devoted to its study. It was *the* scientific study of Archibald Smith; and Airy and Thomson have added not a little to our practical knowledge of the disturbance of the compass by the iron of the ship. Sir W. Thomson, in addition to much valuable practical work on the compass, and experimental work on magnetism, has given the most complete and elegant mathematical theory of the subject*. Of late years the development of the dynamo machine has directed attention to the magnetisation of iron from a different point of view, and a very great deal has been done by many workers to ascertain the facts regarding the magnetic properties of iron. The upshot of these many years of study by practical men interested in the mariner's compass, or in dynamo machines by theoretical men interested in looking into the nature of things, is, that although we know a great many facts about magnetism, and a great deal about the relation of these facts to each other, we are as ignorant as ever we were as to any reason why the earth is a magnet, as to why its magnetic poles are in slow motion in relation to its substance, or as to why iron, nickel, and cobalt are magnetic, and nothing else, so far as we know, is to any practical extent. In most branches of science the more facts we know the more fully we recognise a continuity in virtue of which we see the same property running through all the various forms of matter. It is not so in magnetism; here the more we know the more remarkably exceptional does the property appear, the less chance does there seem to be of resolving it into anything else. It seems to me that I cannot better occupy the present occasion than by recalling your attention to, and inviting discussion of, some of those salient properties of magnetism as exhibited by iron, nickel, and cobalt—properties most of them very familiar, but properties which any theory of magnetism must reckon with and explain. We shall not touch on the great subject of the earth as a magnet—though much has been recently done, particularly by Rücker and Thorpe—but deal simply with magnetism as a property of these three bodies, and consider its natural history, and how it varies with the varying condition of the material.

To fix our ideas, let us consider, then, a ring of uniform section of any convenient area and diameter. Let us suppose this ring to be wound with copper wire, the convolutions being insulated. Over the copper wire let us suppose that a second wire is wound, also insulated, the coils of each wire being arranged as are the coils of any ordinary modern transformer. Let us suppose that the ends of the inner coil, which we will call the secondary coil, are connected to a ballistic galvanometer;

* *Papers on Electrostatics and Magnetism*, p. 340 *et seq.*

and that the ends of the outer coil, called the primary, are connected, through a key for reversing the current, with a battery. If the current in the primary coil is reversed, the galvanometer needle is observed to receive a sudden or impulsive deflection, indicating that for a short time an electromotive force has been acting on the secondary coil. If the resistance of the secondary circuit is varied, the sudden deflection of the galvanometer needle varies inversely as the resistance. With constant resistance of the secondary circuit the deflection varies as the number of convolutions in the secondary circuit. If the ring upon which the coils of copper wire are wound is made of wood or glass—or, indeed, of 99 out of every 100 substances which could be proposed—we should find that for a given current in the primary coil the deflection of the galvanometer in the secondary circuit is substantially the same. The ring may be of copper, of gold, of wood, or glass,—it may be solid or it may be hollow,—it makes no difference in the deflection of the galvanometer. We find, further, that with the vast majority of substances the deflection of the galvanometer in the secondary circuit is proportional to the current in the primary circuit. If, however, the ring be of soft iron, we find that the conditions are enormously different. In the first place, the deflections of the galvanometer are very many times as great as if the ring were made of glass, or copper, or wood. In the second place, the deflections on the galvanometer in the secondary circuit are not proportional to the current in the primary circuit; but as the current in the primary circuit is step by step increased we find that the galvanometer deflections increase somewhat as is illustrated in the accompanying curve (Fig. 1), in which the abscissæ are proportional to the primary current, and the ordinates are proportional to the galvanometer deflections. You observe that as the primary current is increased the galvanometer deflection increases at first at a certain rate; as the primary current attains a certain value the rate at which the deflection increases therewith is rapidly increased, as shown in the upward turn in the curve. This rate of increase is maintained for a time, but only for a time. When the primary current attains a certain value the curve bends downward, indicating that the deflections of the galvanometer are now increasing less rapidly as the primary current is increased; if the primary current be still continually increased, the galvanometer deflections increase less and less rapidly.

Now what I want to particularly impress upon you is the enormous difference which exists between soft iron on the one hand, and ordinary substances on the other. On this diagram I have taken the galvanometer deflections to the same scale for iron, and for such substances as glass or wood. You see that the deflections in the case of glass or wood, to the same scale, are so

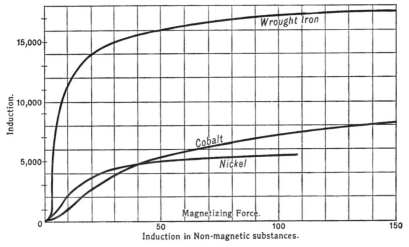

Fig. 1.

small as to be absolutely inappreciable, whilst the deflection for iron at one point of the curve is something like 2000 times as great as for non-magnetic substances. This extraordinary property is possessed by only two other substances besides iron—cobalt and nickel. On the same figure are curves showing on the same scale what would be the deflections for cobalt and nickel, taken from Professor Rowland's paper[*]. You observe that they show the same general characteristics as iron, but in a rather less degree. Still, it is obvious that these substances may be broadly classed with iron in contradistinction to the great mass of other bodies. On the other hand, diamagnetic bodies belong distinctly to the other class. If the deflection with a non-magnetic ring be unity, that with iron, as already stated, may be as much as 2000; that with bismuth, the most powerful diamagnetic known, is 0·999825 —a quantity differing very little from unity. Note, then, the first fact which any theory of magnetism has to explain is: Iron,

[*] *Phil. Mag.*, Nov. 1874.

nickel, and cobalt, all enormously magnetic; other substances practically non-magnetic. A second fact is: With most bodies the action of the primary current on the secondary circuit is strictly proportional to the primary current; with magnetic bodies it is by no means so.

You will observe that the ordinates in these curves, which are proportional to the kicks or elongations of the galvanometer, are called induction, and that the abscissæ are called magnetising force. Let us see a little more precisely what we mean by the terms, and what are the units of measurement taken. The elongation of the galvanometer measures an impulsive electromotive force—an electromotive force acting for a very short time. Charge a condenser to a known potential, and discharge it through the galvanometer: the needle of the galvanometer will swing aside through a number of divisions proportional to the quantity of electricity in the condenser—that is, to the capacity and the potential. From this we may calculate the quantity of electricity required to give a unit elongation. Multiply this by the actual resistance of the secondary circuit and we have the impulsive electromotive force in volts and seconds, which will, in the particular secondary circuit, give a unit elongation. We must multiply this by 10^8 to have it in absolute C.G.S. units. Now the induction is the impulsive electromotive force in absolute C.G.S. units divided by the number of secondary coils and by the area of section of the ring in square centimetres. The line integral of magnetising force is the current in the primary in absolute C.G.S. units—that is, one-tenth of the current in ampères —multiplied by 4π and by the number of convolutions in the primary coil. The magnetising force is the line integral divided by the length of the line over which that line integral is distributed. This is, in truth, not exactly the same for all points of the section of the ring—an imperfection so far as it goes in the ring method of experiment. The absolute electro-magnetic C.G.S. units have been so chosen that if the ring be perfectly non-magnetic the induction is equal to the magnetising force. We may refer later to the permeability, as Sir W. Thomson calls it; it is the ratio of the induction to the magnetising force causing it, and is usually denoted by μ.

There is a further difference between the limited class of magnetic bodies and the great class which are non-magnetic.

To show this, we may suppose our experiment with the ring to be varied in one or other of two or three different ways. To fix our ideas, let us suppose that the secondary coil is collected in one part of the ring, which, provided that the number of turns in the secondary is maintained the same, will make no difference in the result in the galvanometer. Let us suppose, further, that the ring is divided so that its parts may be plucked from together, and the secondary coil entirely withdrawn from the ring. If now the primary current have a certain value, and if the ring be plucked apart and the secondary coil withdrawn, we shall find that, whatever be the substance of which the ring is composed, the galvanometer deflection is one-half of what it would have been if the primary current had been reversed. I should perhaps say approximately one-half, as it is not quite strictly the case in some samples of steel, although, broadly speaking, it is one-half. This is natural enough, for the exciting cause is reduced from—let us call it a positive value, to nothing when the secondary coil is withdrawn; it is changed from a positive value to an equal and opposite negative value when the primary current is reversed. Now comes the third characteristic difference between the magnetic bodies and the non-magnetic. Suppose that, instead of plucking the ring apart when the current had a certain value, the current was raised to this value and then gradually diminished to nothing, and that then the ring was plucked apart and the secondary coil withdrawn. If the ring be non-magnetic, we find that there is no deflection of the galvanometer; but, on the other hand, if the ring be of iron, we find a very large deflection, amounting, it may be, to 80 or 90 per cent. of the deflection caused by the withdrawal of the coil when the current had its full value. Whatever be the property that the passing of the primary current has imparted to the iron, it is clear that the iron retains a large part of this property after the current has ceased. We may push the experiment a stage further. Suppose that the current in the primary is raised to a great value, and is then slowly diminished to a smaller value, and that the ring is opened and the secondary coil withdrawn. With most substances we find that the galvanometer deflection is precisely the same as if the current had been simply raised to its final value. It is not so with iron: the galvanometer deflection depends not alone upon the current at the moment of withdrawal,

MAGNETISM.

but on the current to which the ring has been previously subjected. We may then draw another curve (Fig. 2) representing the galvanometer deflections produced when the current has been

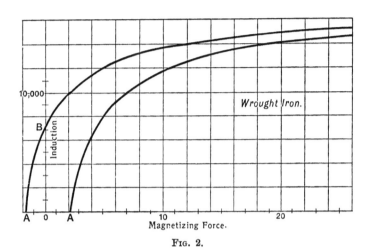

Fig. 2.

raised to a high value and has been subsequently reduced to a value indicated by the abscissæ. This curve may be properly called a descending curve. In the case of ordinary bodies this curve is a straight line coincident with the straight line of the ascending curve, but for iron is a curve such as is represented in the drawing. You observe that this curve descends to nothing like zero when the current is reduced to zero; and that when the current is not only diminished to zero, but is reversed, the galvanometer deflection only becomes zero when the reversed current has a substantial value. This property possessed by magnetic bodies of retaining that which is impressed upon them by the primary current has been called by Professor Ewing "hysteresis[*]," or, as similar properties have been observed in quite other connections, "magnetic hysteresis." The name is a good one, and has been adopted. Broadly speaking, the induction as measured by the galvanometer deflection is independent of the time during which the successive currents have acted, and depends only upon their magnitude and order of succession. Some recent experiments of Professor Ewing, however, seem to show a well-

---

[*] *Phil. Trans. R. S.*, 1885, Part II. p. 526.

marked time effect*. There are curious features in these experiments which require more elucidation.

It has been pointed out by Warburg, and subsequently by Ewing†, that the area of curve 2 is a measure of the quantity of energy expended in changing the magnetism of the mass of iron from that produced by the current in one direction to that produced by the current in the opposite direction and back again. The energy expended with varying amplitude of magnetising forces has been determined for iron, and also for large magnetising forces for a considerable variety of samples of steel‡. Different sorts of iron and steel differ from each other very greatly in this respect. For example, the energy lost in a complete cycle of reversals in a sample of Whitworth's mild steel was about 10,000 ergs per cubic centimetre; in oil-hardened hard steel it was near 100,000; and in tungsten steel it was near 200,000—a range of variation of 20 to 1. It is, of course, of the greatest possible importance to keep this quantity low in the case of armatures of dynamos, and in that of the cores of transformers. If the armature of a dynamo machine be made of good iron, the loss from hysteresis may easily be less than 1 per cent.; if, however, to take an extreme case, it were made of tungsten steel, it would readily amount to 20 per cent. In the case of transformers and alternate-current dynamo machines, where the number of reversals per second is great, the loss of power by hysteresis of the iron, and the consequent heating, become very important. The loss of power by hysteresis increases more rapidly than does the induction. Hence it is not well in such machines to work the iron to anything like the same intensity of induction as is desirable in ordinary continuous-current machines. The quantity OA when measured in proper units, as already explained—that is to say, the reversed magnetic force, which just suffices to reduce the induction as measured by the kick on the galvanometer to nothing after the material has been submitted to a very great magnetising force—is called the "coercive force," giving a definite meaning to a term which has long been used in a somewhat indefinite sense. The quantity is really the important one in judging the magnetism of

---

* Paper by Professor Ewing, "On Hysteresis in the Relation of Strain to Stress," read before the British Association, Sept. 1889.

† *Report Brit. Assoc.* 1883.

‡ Hopkinson, *Phil. Trans.* 1885, Part II.

short permanent magnets. The residual magnetism, OB, is then practically of no interest at all; the magnetic moment depends almost entirely upon the coercive force. The range of magnitude is somewhat greater than in the case of the energy dissipated in a complete reversal. For very soft iron the coercive force is 1·6 C.G.S. units; for tungsten steel, the most suitable material for magnets, it is 51 in the same units. A very good guess may be made of the amount of coercive force in a sample of iron or steel by the form of the ascending curve, determined as I described at first. This is readily seen by inspection of Fig. 3, which shows

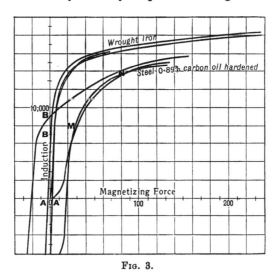

Fig. 3.

the curves in the cases of wrought-iron, and steel containing 0·9 per cent. of carbon. With the wrought-iron a rapid ascent of the ascending curve is made, when the magnetising force is small and the coercive force is small; in the case of the hard steel the ascent of the curve is made with a larger magnetising current, and the coercive force is large. There is one curious feature shown in the curve for hard steel which may, so far as I know, be observed in all magnetisable substances: the ascending curve twice cuts the continuation of the descending curve, as at M and N. This peculiarity was, so far as I know, first observed by Professor G. Wiedemann.

I have already called emphatic attention to the fact that magnetic substances are enormously magnetic, and that non-magnetic substances are hardly at all magnetic; there is between the two classes no intermediate class. The magnetic property of iron is exceedingly easily destroyed. If iron be alloyed with 12 per cent. of manganese, the kick on the galvanometer which the material will give, if made into a ring, is only about 25 per cent. greater than is the case with the most completely non-magnetic material, instead of being some hundreds of times as great, as would be the case with iron*. Further, with this manganese steel, the kick on the galvanometer is strictly proportional to the magnetising current in the primary, and the material shows no sign of hysteresis. In short, all its properties would be fully accounted for if we supposed that manganese steel consisted of a perfectly non-magnetic material, with a small percentage of metallic iron mechanically admixed therewith. Thus the property of non-magnetisability of manganese steel is an excellent proof of the fact—which is also shown by the non-magnetic properties of most compounds of iron—that the property appertains to the molecule, and not to the atom; or to put it in another way, suppose that we were to imagine manganese steel broken up into small particles, as these particles became smaller there would at length arrive a point at which the iron and the manganese would be entirely separated from each other: when this point is reached the particles of iron are non-magnetic. By the magnetic molecule of the substance we mean the smallest part which has all the magnetic properties of the mass. The magnetic molecule must be big enough to contain its proportion of manganese. In iron, then, we must have a collection of particles of such magnitude that it would be possible for the manganese to enter into each of them, to constitute an element of the magnet. Manganese is, so far as I know, a non-magnetic element. Smaller proportions of manganese reduce the magnetic property in a somewhat less degree, the reduction being greater as the quantity of manganese is greater. It appeared very possible that the non-magnetic property of manganese steel was due to the coercive force being very great—that, in fact, in all experiments we were still on that part of the magnetisation curve below the rapid rise, and that if the steel were submitted to greater forces it would presently prove to be magnetic, like other kinds of steel. Professor Ewing, however, has submitted manganese steel to very great forces

* Hopkinson, *Phil. Trans.* 1885, Part II. p. 462.

222 MAGNETISM.

indeed, and finds that its magnetism is always proportional to the magnetising force.

No single body is known having the property of capacity for magnetism in a degree which is neither very great nor very small, but intermediate between the two extremes. We can, however, mix magnetic and non-magnetic substances to form bodies apparently intermediate. It is, therefore, interesting to consider what the properties might be of such a mixture. It depends quite as much on the way in which the magnetic part is arranged in the mass, as on its actual quantity. Suppose, for example, it is arranged as in Fig. 4—in threads or plates having a very long

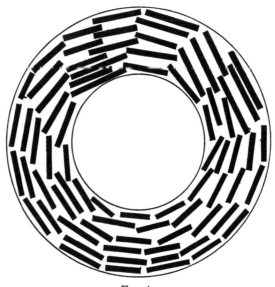

Fig. 4.

axis in the direction of the magnetising force—we may at once determine the curve of magnetisation of the mixture from that of the magnetic substance by dividing the induction for any given force in the ratio of the whole volume to the volume of magnetic substance. If, on the other hand, it is as in Fig. 5—with a very short axis in the direction of the force, and a long axis perpendicular thereto—we can equally construct the curve of magnetisation. This is done in Fig. 6, which shows the curve when nine-tenths of the material is highly magnetic iron arranged as in fig. 5, whilst the other curve of the same figure is that when only one-tenth is magnetic, but arranged as in Fig. 4. You observe how very different is the character of the curve—a difference which is

Fig. 5.

reduced by the much less proportion of magnetic material in the mixture in the one case than in the other. One peculiarity of

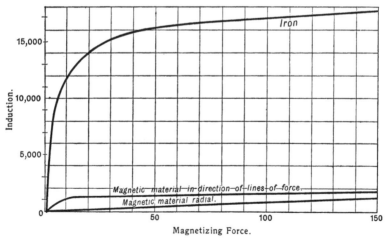

Fig. 6.

these arrangements of the two materials in relation to each other is, that the resulting material is not isotropic; that is, its properties are not the same in all directions, but depend upon the direction of the magnetising force in the material. Of course, this is not at all a probable arrangement, but it is instructive in showing the character of the result as depending upon the construction of the material. Let us, however, consider the simplest isotropic arrangement; let us suppose that one material is in the form of spheres bedded in a matrix of the other: if the spheres are placed at random this is clearly an isotropic arrangement. The result is very different according as the matrix or the spheres are of the magnetic material. Suppose that the volume of the spheres is one-half of the whole volume. In Fig. 7 we have approximately the curve for iron, for a mixture

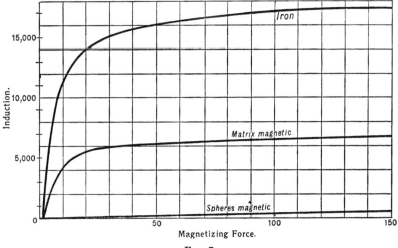

Fig. 7.

of equal quantities of iron and a non-magnetic material; the spheres being non-magnetic and the matrix iron, and for a mixture, the spheres being iron, and the matrix non-magnetic. Observe the great difference. When the spheres are iron, the induction is near four times the force for all values of the force. When the matrix is iron, the induction is near ⅔ths of the induction when the material is iron only.

In speaking of the properties of bodies which, like manganese steel, are slightly magnetic, it may be well here to enter a caution. But little that is instructive is to be learned by testing filings, or the like, with magnets, as these show but little difference between bodies which are slightly magnetic and those which are strongly magnetic. Suppose the filings to be spheres, in the following table are given comparative values of the forces they would experience in terms of μ, if placed in a magnetic field of given value, μ having its ordinary signification—that is, being the ratio of the kick on the galvanometer when a ring is tried made of the material of the filing to the kick if the ring is made of a perfectly non-magnetic material:—

μ.	Attraction.	
1	0	Non-magnetic body.
1·47	0·18	Manganese steel with 12 °/$_o$.
3·6	1·2	Manganese steel with 9 °/$_o$.
5	1·5	
10	2·1	
100	2·8	
1,000	2·98.	

Now bodies in which μ is so small as 3·6 belong distinctly to the non-magnetic class; but the test with the magnet would very markedly distinguish them from manganese steel with 12 per cent. of manganese. The distinction, however, between $\mu = 3·6$ and $\mu = 1,000$ is comparatively small; whereas, under the conditions of experiment, μ is much more than 1,000 for most bodies of which iron is the principal constituent.

The effect of stress on the magnetic properties of iron and nickel have been studied by Sir W. Thomson*. A fact interesting from a broad and general point of view is that the effects of stress are different in kind in the case of iron and nickel. In the case of iron, for small magnetising forces in the direction of the tension, tension increases the magnetisation; for large forces, diminishes it. In the case of nickel the effect is always to diminish the magnetisation.

When one considers that the magnetic property is peculiar to three substances,—that it is easily destroyed by the admixture of some foreign body, as manganese,—one would naturally expect that its existence would depend also on the temperature of the body. This is found to be the case. It has long been known that iron remains magnetic to a red heat, and that then it some-

---

* *Mathematical and Physical Papers*, vol. II. p. 332.

what suddenly ceases to be magnetic, and remains at a higher temperature non-magnetic. It has long been known that the same thing happens with cobalt, the temperature of change, however, being higher; and with nickel, the temperature being lower. The magnetic characteristics of iron at a high temperature are interesting*. Let us return to our ring, and let us suppose that the coils are insulated with a refractory material, such as asbestos paper, and that the ring is made of the best soft iron. We are now in a position to heat the ring to a high temperature, and to experiment upon it at high temperatures in exactly the same way as before. The temperature can be approximately determined by the resistance of one of the copper coils. Suppose, first, that the current in the primary circuit which we use for magnetising the ring is small; that from time to time, as the ring is heated and the temperature rises, an experiment is made by reversing the current in the primary circuit and observing the deflection of the galvanometer needle. At the ordinary temperature of the air the deflection is comparatively small; as the temperature increases the deflection also increases, but slowly at first; when the temperature, however, reaches something like 600° C., the galvanometer deflection begins very rapidly to increase, until, with a temperature of 770° C., it attains a value of no less than 11,000 times as great as the deflection would be if the ring had been made of glass or copper, and the same exciting current had been used. Of course a direct comparison of 11,000 to 1 cannot be made: to make it we must introduce resistance into the secondary circuit when the iron is used; and we must, in fact, make use of larger currents when copper is used. However, the ratio of the induction in the case of iron to that in the case of copper, at 770° C., for small forces is no less than 11,000 to 1. Now mark what happens. The temperature rises another 15° C.: the deflection of the needle suddenly drops to a value which we must regard as infinitesimal in comparison to that which it had at a temperature of 770° C.; in fact, at the higher temperature of 785° C. the deflection of the galvanometer with iron is to that with copper in a ratio not exceeding that of 1·14 to 1. Here, then, we have a most remarkable fact: at a temperature of 770° C. the magnetisation of iron 11,000 times as great as that of a non-magnetic substance; at a temperature of 785° C. iron practically

* Hopkinson, *Phil. Trans.* 1889.

non-magnetic. These changes are shown in Fig. 8. Suppose now that the current in the primary circuit which serves to magnetise

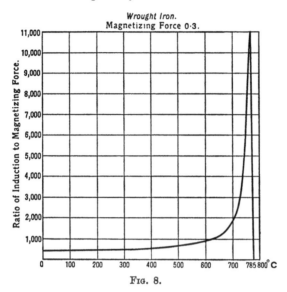

Fig. 8.

the iron had been great instead of very small. In this case we find a very different order of phenomena. As the temperature rises the deflection on the galvanometer diminishes very slowly till a high temperature is attained; then the rate of decrease is accelerated until, as the temperature at which the sudden change occurred for small forces is reached, the rate of diminution becomes very rapid indeed, until, finally, the magnetism of the iron disappears at the same time as for small forces. Instead of following the magnetisation with constant forces for varying temperatures, we may trace the curve of magnetisation for varying forces with any temperature we please. Such curves are given in Diagrams 9 and 10. In the one diagram, for the purpose of bringing out different points in the curve, the scale of abscissæ is 20 times as great as in the other. You will observe that the effect of rise of temperature is to diminish the maximum magnetisation of which the body is capable, slowly at first, and rapidly at the end. It is also very greatly to diminish the coercive force, and to increase the facility with which the body is magnetised[*].
To give an idea of the magnetising forces in question, the force

---

[*] *Vide* Wiedemann, *Elektricität*, vol. III. p. 769.

228   MAGNETISM.

for Fig. 8 was 0·3; and as you see from Figs. 9 and 10, the force ranges as high as 60. Now the earth's force in these latitudes is 0·43, and the horizontal component of the earth's force is 0·18. In

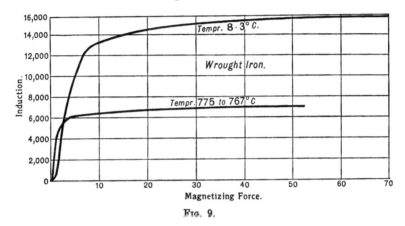

Fig. 9.

the field of a dynamo machine the force is often more than 7,000. In addition to the general characteristics of the curve of magnetisation, a very interesting, and, as I take it, a very important, fact comes out. I have already stated that if the ring be submitted to a great current in one direction, which current is afterwards gradually reduced to zero, the ring is not in its non-magnetic

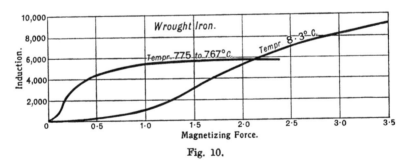

Fig. 10.

condition, but that it is, in fact, strongly magnetised. Suppose now we heat the ring, whilst under the influence of a strong magnetising current, beyond the critical temperature at which it ceases to have any magnetic properties, and that then we reduce the current to zero, we may in this state try any experiment we please. Reversing the current on the ring, we shall find that it is in all cases non-magnetic. Suppose next that we allow the ring to cool without any current in the primary, when cold we find that the ring is magnetised; in fact, it has a distinct recollection of what had been done to it before it was heated to the temperature at which it ceased to be magnetic. When steel is tried in the same way with varying temperatures, a similar sequence of phenomena is observed; but for small forces the permeability rises to a lower maximum, and its rise is less rapid. The critical temperature at which magnetism disappears changes rapidly with the composition of the steel. For very soft charcoal iron wire the critical temperature is as high as 880° C.; for hard Whitworth steel it is 690° C.

The properties of an alloy of manganese and iron are curious. More curious are those of an alloy of nickel and iron[*]. The alloy of nickel and iron containing 25 per cent. of nickel is non-magnetic as it is sure to come from the manufacturer; that is to say, a substance compounded of two magnetic bodies is non-magnetic. Cool it, however, a little below freezing, and its properties change: it becomes very decidedly magnetic. This is perhaps not so very remarkable: the nickel steel has a low critical temperature—lower than we have observed in any other magnetisable body. But if now the cooled material be allowed to return to the ordinary temperature it is magnetic; if it be heated it is still magnetic, and remains magnetic till a temperature

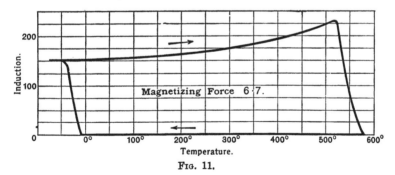

Fig. 11.

of 580° C. is attained, when it very rapidly becomes non-magnetic, exactly as other magnetic bodies do when they pass their critical temperature. Now cool the alloy: it is non-magnetic, and remains non-magnetic till the temperature has fallen to below freezing. The history of the material is shown in Fig. 11, from which it will

[*] Hopkinson, *Proc. R. S.* 1889.

be seen that from $-20°$ C. to $580°$ C. this alloy may exist in either of two states, both quite stable—a magnetic and a non-magnetic—and that the state is determined by whether the alloy has been last cooled to $-20°$ C. or heated to $580°$ C.

Sudden changes occur in other properties of iron at this very critical temperature at which its magnetism disappears. For example, take its electrical resistance*. On the curve, Fig. 12,

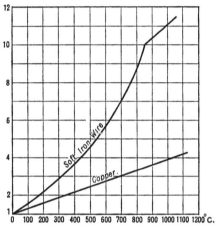

Fig. 12.

is shown the electrical resistance of iron at various temperatures, and also the electrical resistance of copper or other pure metal. Observe the difference. If the iron is heated, its resistance increases with an accelerating velocity, until, when near the critical temperature, the rate of increase is five times as much as the copper; at the critical temperature the rate suddenly changes, and it assumes a value which, as far as experiments have gone, cannot be said to differ very materially from a pure metal. The resistance of manganese steel shows no such change; its temperature coefficient constantly has the value of ·0012, which it has at the ordinary temperature of the air. The electrical resistance of nickel varies with temperature in an exactly similar manner. Again, Professor Tait has shown that the thermo-electric properties of iron are very anomalous—that there is a sudden change at or about the temperature at which the metal becomes

* Kohlrausch, *Wiedemann's Annalen*, 1888; Hopkinson, *Phil. Trans. R. S.* 1889.

non-magnetic, and that before this temperature is reached the variations of thermo-electric property are quite different from a non-magnetic metal.

Professor Tomlinson has investigated how many other properties of iron depend upon the temperature. But the most significant phenomenon is that indicated by the property of recalescence. Professor Barrett, of Dublin, observed that if a wire of hard steel is heated to a very bright redness, and is then allowed to cool, the wire will cool down till it hardly emits any light at all, and that then it suddenly glows out quite bright again, and afterwards finally cools. This phenomenon is observed with great difficulty in the case of soft iron, and is not observed at all in the case of manganese steel. A fairly approximate numerical measurement may be made in this way*: Take a block of iron or steel on which a groove is cut, and in this groove wind a coil of copper wire insulated with asbestos; cover the coil with many layers of asbestos; and finally cover the whole lump of iron or steel with asbestos again. We have now a body which will heat and cool comparatively slowly, and which will lose its heat at a rate very approximately proportional to the difference of temperature between it and the surrounding air. Heat the block to a bright redness, and take it out of the fire and observe the resistance of the copper coil as the temperature falls, due to the cooling of the block. Plot a curve in which the abscissæ are the times, and the ordinates the logarithms, of the increase of resistance of the copper coil above its resistance at the temperature of the room. If the specific heat of the iron were constant, this curve would be a straight line; if at any particular temperature latent heat were liberated, the curve would be horizontal so long as the heat was being liberated. If now a block be made of manganese steel, it is found that the curve is very nearly a straight line, showing that there is no liberation of latent heat at any temperature. If it is made of nickel steel with 25 per cent. of nickel, in its non-magnetic state, the result is the same—no sign of liberation of heat. If now the block be made of hard steel, the temperature diminishes at first; then the curve (Fig. 13) which represents the temperature bends round: the temperature actually rises many degrees whilst the body is losing heat. The liberation of heat

* *Phil. Trans. R. S.* 1889, p. 463.

being completed, the curve finally descends as a straight line. From inspection of this curve it is apparent why hard steel exhibits a sudden accession of brightness as it yields up its heat. In the case of soft iron the temperature does not actually rise as the body loses heat, but the curve remains horizontal, or nearly horizontal, for a considerable time. This, again, shows why, although a considerable amount of heat is liberated at a temperature corresponding to the horizontal part of the curve, no marked recalescence can be obtained. From curves such as these it is easy to calculate the amount of heat which becomes latent. As the iron passes the critical point it is found to be about 200 times as much heat as is required to raise the temperature of the iron 1 degree centigrade. From this we get a very good idea of the

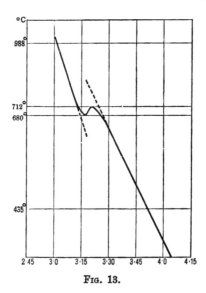

Fig. 13.

importance of the phenomenon. When ice is melted and becomes water, the heat absorbed is 80 times the heat required to raise the temperature of the water 1 degree centigrade, and 160 times the heat required to raise the temperature of the ice by the same amount. The temperature of recalescence has been abundantly identified with the critical temperature of magnetism[*]. I am

[*] I have only recently become acquainted with the admirable work of M. Osmond (*Transformation du Fer et du Carbone*, 1888) on recalescence. He has examined a great variety of samples of steel, and determined the temperatures at which they

not aware that anything corresponding with recalescence has been observed in the case of nickel. Experiments have been tried*, and gave a negative result, but the sample was impure; and the result may, I think, be distrusted as an indication of what it would be in the case of pure nickel. The most probable explanation in the case of iron, at all events, appears to be that when iron passes from the magnetic to the non-magnetic state it experiences a change of state of comparable importance with the change from the solid to the liquid state, and that a large quantity of heat is absorbed in the change. There is, then, no need to suppose chemical change; the great physical fact accompanying the absorption of heat is the disappearance of the capacity for magnetisation.

What explanations have been offered of the phenomena of magnetism? That the explanation must be molecular was early apparent. Poisson's hypothesis was that each molecule of a magnet contained two magnetic fluids, which were separated from each other under the influence of magnetic force. His theory explained the fact of magnetism induced by proximity to magnets, but beyond this it could not go. It gave no hint that there was a limit to the magnetisation of iron—a point of saturation; but slight of hysteresis; no hint of any connection between the magnetism of iron and any other property of the substance; no hint why magnetism disappears at a high temperature. It does, however, give more than a hint that the permeability of iron could not exceed a limit much less than its actual value, and that it should be constant for the material, and independent of the force applied. Poisson gave his theory a beautiful mathematical development, still useful in magnetism and in electrostatics.

give off an exceptional amount of heat. Some of his results are apparent on my own curves, though I had assumed them to be mere errors of observation. For example, referring to my Royal Society paper, comparing Figs. 38 and 38A, we see that the higher the heating, the lower is the point of recalescence; both features are brought out by M. Osmond. The double recalescence observed by M. Osmond in steel with a moderate quantity of carbon I would explain provisionally by supposing this steel to be a mixture of two kinds which have different critical temperatures. Although M. Osmond's method is admirable for determining the temperature of recalescence, and whether it is a single point or multiple, it is not adapted to determine the quantity of heat liberated, as the small sample used is enclosed in a tube of considerable mass, which cools down at the same time as the sample experimented upon.

* *Proc. R. S.* vol. XLIV. 1888, p. 319.

Weber's theory is a very distinct advance on Poisson's. He supposed that each molecule of iron was a magnet with axes arranged at random in the body; that under the influence of magnetising force the axes of the little magnets were directed to parallelism in a greater degree as the force was greater. Weber's theory thoroughly explains the limiting value of magnetisation, since nothing more can be done than to direct all the molecular axes in the same direction. As modified by Maxwell, or with some similar modification, it gives an account of hysteresis, and of the general form of the ascending curve of magnetisation. It is also very convenient for stating some of the facts. For example, what we know regarding the effect of temperature may be expressed by saying that the magnetic moment of the molecule diminishes as the temperature rises, hence that the limiting moment of a magnet will also diminish; but that the facility with which the molecules follow the magnetising force is also increased, hence the great increase of μ for small forces, and its almost instantaneous extinction as the temperature rises. Again, in terms of Weber's theory, we can state that rise of temperature enough to render iron non-magnetic will not clear it of residual magnetism. The axes of the molecules are brought to parallelism by the force which is impressed before and during the time that the magnetic property is disappearing; they remain parallel when the force ceases, though, being now non-magnetic, their effect is *nil*. When, the temperature falling, they become again magnetic, the effect of the direction of their axes is apparent. But Weber's theory does not touch the root of the matter by connecting the magnetic property with any other property of iron, nor does it give any hint as to why the moment of the molecule disappears so rapidly at a certain temperature.

Ampère's theory may be said to be a development of Weber's: it purports to state in what the magnetism of the molecule consists. Associated with each molecule is a closed electric current in a circuit of no resistance; each such molecule, with its current, constitutes Weber's magnetic molecule, and all that it can do they can do. But the great merit of the theory—and a very great one it is—is that it brings magnetism in as a branch of electricity; it explains why a current makes a magnetisable body magnetic. It also gives, as extended by Weber, an explanation of diamagnetism. It, however, gives no hint of connecting the magnetic properties of iron with any other property. Another difficulty is this: When iron ceases to be magnetisable, we must assume that the molecular currents cease. These currents represent energy. We should therefore expect that, when iron ceased to be magnetic by rise of temperature, heat would be liberated; the reverse is the fact.

So far as I know, nothing that has ever been proposed even attempts to explain the fundamental anomaly, Why do iron, nickel, and cobalt possess a property which we have found nowhere else in nature? It may be that at lower temperatures other metals would be magnetic, but of this we have at present no indication. It may be that, as has been found to be the case with the permanent gases, we only require a greater degree of cold to extend the rule to cover the exception*. For the present, the magnetic properties of iron, nickel, and cobalt stand as exceptional as a breach of that continuity which we are in the habit of regarding as a well-proved law of nature.

* Faraday was strongly of opinion that other metals must be magnetisable at a low temperature: *vide Experimental Researches*, vol. II. p. 217 *et seq.*, and vol. III. p. 444.

15.

INAUGURAL ADDRESS. INSTITUTION OF ELECTRICAL ENGINEERS, 16TH JANUARY, 1896.

WHEN it was suddenly proposed to me that I should a second time be your President, I felt considerable reluctance to accept the duty. This reluctance arose from two causes. Of the first I am ashamed—pure laziness—and of it the less I say the better. The second was a *bona fide* doubt as to whether I was specially suitable for the post. But it soon struck me that that was a matter for which those who proposed me were responsible, and not I who accepted, but had not sought the office. But, if I felt a temporary reluctance, do not suppose that I for a moment failed to appreciate the honour which had been done me. To be called by his fellow professional men to fill the chair which has been occupied by men of such distinction as Kelvin, Siemens, Abel, Crookes, not to mention others, is an honour of which anyone may confess himself proud. After once having occupied this chair, to be called to it a second time is a greater honour, for it shows that one's efforts, however imperfect, have been generously appreciated by those for whom they were made. I thank you for the honour, and assure you I will do my best to verify, if I can, the judgment of those who proposed and you who elected me, rather than my own misgivings.

The century which is now just drawing to a close has been a time remarkable above all other times for the extraordinary development of our knowledge of the physical sciences, and in particular of experimental physics and chemistry, and for the extent of the practical application of this knowledge to useful purposes. The century has seen the discovery of the mechanical nature of heat, of the spectroscope, and of the whole of the science of organic chemistry; it has seen the enormous development of the application of the steam-engine, the construction of our railway system, and the supersession of sailing vessels by steamboats. But perhaps the most remarkable illustration of this development is to be found in the science, and its application, with which this Institution has to deal. Consider what was known of the sciences of electricity and magnetism at the end of the last century. The two sciences were then quite unconnected. In magnetism the properties of permanent magnets were known— that is to say, it was known that steel needles could be rendered magnetic by rubbing with a loadstone, and that when so rendered magnetic they would tend to remain in that condition; it was known that, if a magnet were broken in parts, each part would exhibit North and South poles like the magnet from which it was broken; it was also known that the earth exhibited properties of magnetism, inasmuch as a magnetic needle free to move would point, on the whole, to the North; broadly speaking, what we know to-day about permanent magnets was known, and that was about all. The practical applications were limited to the use of a magnetic needle in the mariner's compass. The knowledge of electricity was confined to what is now known as electrostatics. That certain bodies when rubbed exhibited peculiar properties, attracting and repelling each other, had long been observed; and the facts expressed by the hypothesis of positive and negative electricity were known—namely, that two pieces of amber which had been rubbed repelled each other, two pieces of glass which had been rubbed repelled each other, but a piece of rubbed glass and a piece of rubbed amber attracted each other. In a general way the properties of conductors, as distinguished from insulators, were also known in their relation to charges of frictional electricity. Cavendish had made elaborate investigations into the capacity of condensers, the condensers being usually of glass, and being charged by the old-fashioned frictional means. Franklin had proved the identity of the cause of lightning with the cause of the sparks produced by frictional electricity. Just one hundred years ago Galvani had made his celebrated experiments showing that the muscles of the limb of a frog were disturbed by contact with a piece of copper and a piece of zinc which touched each other; and that was the only fact which up to that time had ever been observed in the enormous group of facts which now constitute our knowledge of the production of electricity by chemical action. Shortly stated, that was the extent of our theoretical and experimental knowledge of electricity. The practical applications of this knowledge were almost *nil.* I can recall nothing excepting the provision of lightning conductors as a protection to buildings.

It is worth insisting that the additions to our knowledge of the properties of permanent magnets and of frictional electricity during the century have not been of great importance, and the practical applications of these limited branches of knowledge have been few. If one is asked what uses can be made of our knowledge of frictional electricity apart from the phenomena of currents, it is hard to name anything of importance. In like manner, in regard to our knowledge of permanent magnets, it may almost be said that the advances in practical applications have been confined to improvements in the use of the mariner's compass, which, most important although they are, have had comparatively little share in the changes in our conditions which the last hundred years have witnessed. Almost everything in the way of practical applications of electrical science has been the outcome of subsequent discoveries—the voltaic cell, discovered by Volta in the year 1800; Davy's discoveries in electrolysis; Oersted's discovery of the action of currents upon magnets; Ampère's discoveries of the mechanical action of currents upon currents; and Faraday's discoveries in electro-magnetism; not to mention the names of eminent men still living. Now the remarkable thing is that none of these phenomena had ever been witnessed by mortal man. It was not that they were constantly occurring around us, and that the explanation had not been seen; it is the fact that the phenomena had never been experienced. It would seem that the phenomena of electric currents as Volta discovered them do not occur in nature on any substantial scale. Electro-magnets may be regarded as purely a creation of the human mind; so far as we know at present, no example of an electro-magnet is found in the natural world. The great currents produced by the action of electro-magnets have, so far as we know, no counterpart in nature—not only no counterpart in degree, but none in kind. In this respect our science differs much from many others. In mechanics and in heat we may find the phenomenon of which we make use occurring without our aid on an enormously greater scale than we are accustomed to use; changes are constantly occurring around us, and it is conceivable that we might have obtained much of our present knowledge by the aid of observation without experiments. But it is not so with electricity and magnetism. We should be puzzled even now to adequately illustrate the facts we know by observation alone without experiment.

Let us consider for a moment the practical applications which have been made of electrical knowledge during the last 60 years. The first, of course, is the telegraph. What are the scientific facts which have been used in the development of telegraphy? The currents are generated either by chemical means or electromagnetic means. The receiving instruments either depend for their action on the action of currents upon magnets, or, in a limited number of cases, on electro-chemical decomposition. We might put it broadly thus—that, with the exception of the distinction between conductors and non-conductors and certain properties of condensers, the telegraph could do without anything that was known up to the end of the last century, and that it wholly depends upon more recent discoveries. Much the same may be said with regard to the more recent applications: the electric light, electric transmission of power, and the rapidly growing department of electro-chemistry may all be said to be quite independent of any facts which were known before the year 1800, and entirely based upon discoveries made since. Suppose it were a fact that electricity could not be produced by friction, that we knew nothing concerning the attraction which oppositely electrified bodies have for each other, that permanent magnets did not exist—or, in other words, that iron had no hysteresis—would it make any material difference, so far as we know, in the practical applications of electrical science to telegraphy, telephony, electric light, electric transmission of power, or electro-chemistry? The permanent magnets, which are practically used only in instruments, would be at once replaced by electro-magnets, and the apparatus would work just as it does now.

Enough has been said, I think, to show that the knowledge of electricity and magnetism which was possessed by men at the end of the last century, intensely interesting though it was, has not been the parent of the great applications of later days; that the knowledge of the last century has not grown in any great degree upon the lines which existed at its commencement, but that our advances in science, as in practice, were the result of the discoveries which were made since the beginning of the present century.

It has been a frequent practice with presidents of societies devoted to promoting some department of science, theoretic or applied, to deal with the history of the science in their presidential addresses,—to show how one idea has led up to another, and how marked have been the advances made. This evening I propose to adopt a different course. What I wish to consider is, how would the theory of electricity have been arranged if the order of discovery of the facts of the science had been other than it has been? Here I owe you some words of apology. You are a practical body more immediately interested in the application of science to the service of man than in the promotion of abstract knowledge. The justification I would offer for my subject is that the science of electricity, whether viewed from the practical or theoretical side, is in a state of extraordinarily rapid advance. The rules of practice of yesterday are of little use to-day, hence more discussion is needed, more publication is needed, than in departments which are in a more stable state. Furthermore, there is no department of applied science in which the connection with theory is so close and intimate. The practical electrical engineer reads eagerly the results of those who, careless whether their work finds application or not, labour to advance knowledge; he attacks these results whether they be expressed in the ordinary speech of men or in mathematical symbols. On the other hand, much of our knowledge of electricity, viewed from the scientific side, has come from those who are concerned with practical application. Nowhere else do we find a closer interdependence of pure science and practice. In what I have to say there will be much that is trite and obvious to the merest tyro in electricity. What I wish to do is to show how the facts of electrostatics can be explained by the facts of current-electricity without hypothesis; and it is necessary to rehearse elementary principles. In a complicated subject like electricity, in which the appeals to our senses are indirect, there are open to us many ways of arranging our ideas, any one of which will give us a consistent account of that which we observe. We may take as our fundamental basis one or other of various groups of facts, and deduce others therefrom. The actual arrangement of ideas on the subject depends in a measure upon the accident of the historical order of discovery. Thus, the fact that actions at a distance between electrified bodies were known many years before the phenomena of current-electricity and electro-magnetism is the reason why that part of the science which treats of such actions is to-day treated first and made the basis of the rest. The time has, I think, come when we may with advantage reconsider our position, and see whether or not our ideas may be more conveniently arranged. The arrangement which is most convenient may no doubt depend upon the phenomena which occupy the most prominent place in the practical use we wish to make of the science. Those facts are best treated as fundamental which are most frequently used, provided they are sufficiently simple. To the practical electrician, are the facts of current-electricity, or of the attractions and repulsions of electrified bodies, more familiar? Surely the transference of energy by current is the agent by which almost every operation he purposes is effected, whilst the phenomena of electrostatics touch him only at points in the theory of electrometers and so forth. I propose, then, this evening to sketch out very briefly, *ab initio*, how we may base the science of electrostatics upon the theory of current-electricity taken as fundamental. We might, if we pleased, drop the word "current" altogether and substitute another. It seems hardly worth while, but I shall ask you to bear in mind that, if used to-night, it does not connote the idea of anything flowing along the conductor. It is a name for a directed magnitude, and nothing more. I shall have occasion to refer to some elementary experiments, but you will bear with me, I trust, for we are concerned with arranging the subject on as small a basis of facts as is consistent with rigid deduction. Please, then, to banish from your minds all theory on the subject, and let us imagine that the facts of current-electricity are first discovered, and that it is only when these are well developed that electrostatic attractions and repulsions are found out.

In a vessel of dilute sulphuric acid is placed a piece of pure zinc, or of zinc carefully amalgamated, and also a piece of copper, platinum, silver, or carbon, the two plates not being in contact with each other. It is observed that no change takes place; the acid does not attack either plate. The two plates are now touched with the two ends of a wire of copper or other metal; immediately the zinc begins to dissolve in the sulphuric acid, and gas—which, if collected, proves to be hydrogen—is found on the copper or platinum plate. If the plates are large and near together, and the wire is not too long or thick, it becomes very sensibly heated. On separating the wire from either plate, the chemical reaction and the heating of the wire cease. When the wire touches both plates there is then development of chemical energy in the vessel of acid; but a part of this energy appears as heat, not where it is developed, but in the wire connecting the plates. Whether this energy goes across the space of air between the zinc plate and the wire, or whether it goes along the wire, we do not know; but that the two plates should be rendered continuous with each other by a wire or some equivalent is a condition precedent to the chemical reaction and to the heating of the wire.

Let two such vessels of dilute acid, with plates of zinc and copper, be provided, and let the copper in one vessel be connected metallically to the zinc in the other: nothing happens until the copper in the second vessel is connected to the zinc of the first; then both zincs are dissolved, hydrogen is liberated on both copper plates, and the connecting wires are more or less heated.

Again, varying the experiment, let the coppers in the two vessels be connected together and the zincs be connected together: no reaction or heating occurs.

The combination of a piece of zinc and a piece of copper dipping separately into a vessel of dilute acid is called a "galvanic cell," or element; a series of elements connected together copper to zinc is called a "galvanic battery." Galvanic elements and batteries take a multitude of forms: the zinc may be replaced by another metal—for example, cadmium—and the copper by platinum; many other liquids than dilute sulphuric acid may be used; but all galvanic batteries have this in common—to cause them to work, the chain or circuit must be completed by a wire touching the zinc at one end of the series and the copper at the other, and then chemical reactions occur in the cells, and heat is liberated in the metal connections. The zinc and copper plates, or their equivalents, at the ends of a galvanic battery are called "poles."

Let two plates of platinum dip into dilute sulphuric acid; connect one plate by means of a wire to the zinc pole of a battery, the other plate to the copper pole. So soon as the connections are made, chemical reactions occur in the cells of the battery, the connecting wires are more or less heated, and, further, a reaction occurs in the vessel in which the platinum plates dip, the water is decomposed, gases are liberated on the two plates; if these are collected, it is found that the gas from the plate connected to the zinc pole is hydrogen, that from the plate connected to the copper is oxygen. The effects cease on detaching either wire from the platinum plate or the pole of the battery, or otherwise interrupting the continuity of the circuit. In this experiment is seen a chemical reaction liberating energy in the cells of the battery, and another chemical reaction absorbing energy in the vessel containing the platinum plates; how the energy passes from one vessel to the other we do not know.

The experiment may be varied. Let two plates of zinc or other metal dip into a solution of salt of zinc or other metal, and let the plates be connected to the poles of a galvanic battery for a time. If the plates be weighed before and after the operation, it is found that the plate which was connected to the zinc pole has gained in weight, and that the plate which has been connected to the copper has lost an equal amount. By weighing such plates, and noting the time the experiment was allowed to continue, it is practicable to examine quantitatively the phenomena of a galvanic circuit. Such pair of plates and solution of salt is called a "voltameter."

Suppose now a series of voltameters, having in each the same metal, are connected together in a chain, a plate of one to a plate of the next, and the two ends of the chain connected to the two poles of a battery, it is found that, no matter what be the size and form of the voltameters, their position, or the arrangement of the connecting wires, there is the same gain and loss of weight of the plates in all the voltameters; it is further found that, if the zincs of the cells of the battery are weighed before and after the experiment, and if the metal used in the voltameter be zinc, the quantity of zinc dissolved is the same in all the cells, and is the same as that transferred in each voltameter. The rate of dissolution and deposit is the same in the several voltameters and cells. In a closed galvanic circuit we have, then, a measurable something the same for all cross-sections of the circuit, measurable by the quantity of zinc or other selected metal which is transferred from one plate to the other of a voltameter is a unit of time. This something has direction as well as magnitude, and as the magnitude may be represented in terms of any unit chosen by a numerical quantity C, so the direction may be represented by the sign $+$ or $-$. It is arbitrarily chosen to take as $+$ the direction from the plate which loses weight to that which gains weight through the liquid in the voltameter. The quantity C taken, with its sign, is called the galvanic or electric current round the circuit, but it must not be supposed that it is known that there is anything flowing round the circuit.

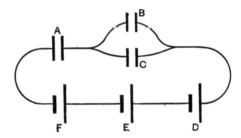

Suppose that, instead of a simple galvanic circuit, the circuit divides or branches as indicated in the sketch, in which A, B, C, are voltameters, FED the battery. It is found that the quantity of zinc moved in B, together with the quantity of zinc moved in C, is equal to the quantity moved in A. This gives us the idea of current in the branch B or C, and we may say that the current in A is the algebraic sum of the currents in B and C.

Not all bodies are competent to complete a galvanic circuit. Thus, if the copper wires which serve to connect the poles of the cells of the batteries and the voltameter are any of them replaced by thread, the phenomena of galvanism do not occur; if one of the wires be severed and a piece of paraffined paper interposed, the phenomena do not occur.

Thus we have a distinction of substances into those which are capable of forming continuously part of a galvanic circuit and those which are not; the former are called "conductors," the latter "non-conductors." The distinction will be found to be really one of degree.

Consider a wire forming part of the galvanic circuit: after a short time no change occurs in the wire; it is heated by the current, and it loses heat by heat conduction, convection, or radiation. It is possible by the methods of calorimetry to measure the total quantity of energy leaving the wire as heat by these agencies; it must be equal to the quantity of energy brought into the wire by galvanic agency.

Again, consider the battery itself. A certain chemical reaction is going on; zinc is being oxidised, and other reactions are occurring. Of the net amount of energy liberated per second by these, some goes to heat the battery, but some also appears in parts of the circuit external to the battery—that is, leaves the battery by galvanic agency. Thus a galvanic current is an instrument which takes energy from certain parts of the circuit and takes it into other parts of the circuit; and when the current is steady the algebraic sum of the quantity of energy taken into all parts of the circuit is *nil*. Denote by EC the quantity of energy per unit of time taken into any section, A, B, of the galvanic circuit, the positive direction of the current being from A to B. If the whole of the circuit be broken into sections, the algebraical sum of the values of EC is *nil*, but the value of C is the same for all sections; hence the algebraic sum of the values of E is *nil*. Let there be any quantity, V_a, referred to each point of the circuit, such that $V_a - V_b = E$, the condition that the sums of the values of E shall be *nil* is fulfilled. E is therefore properly described as the difference of two quantities, one peculiar to A, the other to B; it is called the difference of potential between A and B, and V_a is called the potential at A, though it is only with differences of potential that we are concerned.

Suppose now the circuit divides at A into two or more branches, joining again at B, by our definition the difference of potential between A and B is the sum of the energies taken per unit of time into the conductors A, B, divided by the sum of the currents in the conductors A, B. We have now to show that the energy taken into each of the conductors A, B, is equal to the difference of potential between A and B multiplied by the current in that conductor. This is by no means obvious, but depends on an assumption of an experimental truth, viz., that in a given conductor the energy taken into it by the current is a function of the current only, and is independent of other conditions external to the conductor. Granted this, it easily follows that, if two or more conductors be joined at A and B, the potential difference between A and B deduced from each will be the same.

As we have defined difference of potential by means of work done in a conductor when a given current exists in it, the laws of resistance must have the same basis. In the case of a homogeneous metallic conductor under given constant physical conditions of temperature, stress, and magnetic field carrying a constant galvanic current C, Joule found that the heat liberated per unit of time varies as the square of $C = RC^2$ when R is a constant relative to C for the conductor. Whence follows that $E = RC$. R is called the electrical resistance of the conductor. For metallic conductors R is constant, but the constant R does not exist for all conductors; for the electric arc, for example, the heat generated varies more nearly as C than as C^2, and therefore E is more nearly constant than varying as C.

We have now gained two fundamental ideas—that of galvanic current, and difference of potential. In terms of what units are we to measure them? We have only one unit to choose, because EC must be expressed in terms of our mechanical unit of power. If a unit were to be fixed at this stage, it would be natural to define either a unit of current or of resistance the ratio of $\frac{E}{C}$; for example, we might say the unit of current shall be that current whereby a unit weight of distilled water is decomposed per second; or we might say, let the resistance of a piece of wire of defined shape and material at a definite temperature be the unit of resistance. The units in general use have an electromagnetic basis, but be it observed that all rational systems must be such that EC is expressed in some ordinary mechanical units.

We have supposed that galvanic effects are measured by chemical change, and that the rate of galvanic effect—or, as it is generally called, the electric current—is measured by the rate of chemical change. From a theoretical point of view it would be possible to thus measure electric currents; but for practical purposes other means are adopted—electrometers, galvanometers, electro-dynamometers, and the like. With a description of such instruments and their methods of use I have not the time to concern myself this evening. We must assume that we have the means of measuring C, the galvanic or electric current so called, at any instant, and E, the difference of potential, whether these be great or small, and with any desired accuracy.

We should now be in a position to give a detailed account of galvanic batteries and of electrolysis, and the theory of the distribution of currents in systems of conductors of any form, and we are in a position to place the theory of thermo-electricity upon a thermo-dynamic basis without the introduction of any fresh idea. These, however, I leave on one side, and hasten to consider how the phenomena of electrostatics can be based upon what we have been discussing. There is a broad group of facts very familiar to the practical electrician engaged in laying telegraphic cables of which I have as yet given no account: I refer to the phenomena of capacity. Let us examine a little more closely the statement that I made that if a slip of paraffined paper were interposed between the two parts of a conductor no electrical action would occur. Suppose that, instead of paraffined paper interposed between the ends of a wire, we have a very large number of sheets of tinfoil separated from each other by paraffined paper in such wise that we have a great area of tinfoil connected together, and separated from another great area of tinfoil by nothing more than the paraffined paper. If now these two areas of tinfoil are suddenly connected to the poles of a galvanic battery, we find that it is by no means true that no galvanic effect will ensue; we find that a galvanic current exists, but that, whilst it begins with a large quantity, this quantity goes on diminishing until, barring the slight imperfection of insulation of the paraffined paper, it becomes *nil*. We find, then, that it is not true to state that paraffined paper cannot form a part of a galvanic circuit; it can form a part of a galvanic circuit, but it does so under completely different laws to the metallic conductor. With the metallic conductor the difference of potential at its two ends depends only on the current which then exists in that conductor, and not at all upon the currents which have previously existed; with paraffined paper, on the other hand, the difference of potential between the sheets of tinfoil depends not at all upon the then current, but entirely upon the currents which have existed since its two coatings were connected together, and the times for which they have existed. A galvanic current may be measured by the rate of electrolysis; being measured by a rate, its time integral will have a perfectly definite meaning, and will be measured by the total quantity of substance which would be decomposed in a voltameter in the circuit. We find that the difference of potential between the two coatings of the paraffin condenser is proportional to the time integral of the current since the time at which the coatings were connected together last. The constant ratio of this time integral of current to the difference of potential is called the "capacity" of the condenser. Suppose now that the condenser has been connected to a galvanic battery and the time integral of the current has been measured; let now the coatings of the condenser be disconnected from the battery and be connected together through any instrument for measuring the time integral of current: if this is accomplished with sufficient rapidity for loss by imperfect insulation to be negligible, we find that the time integral in the second case is equal and opposite to that which it was in the first. Here, again, is a difference of conduction in the ordinary sense. If a conductor in which a current has existed be disconnected from the source of current and be connected to any measuring instrument, no return current will be observed. Here, then, we have a new order of facts presented in an extreme case, but, in fact, applicable to all kinds of insulators. Suppose we diminish the area of surface of the paraffined paper, we diminish *pro rata* the capacity of the condenser; but even when the paraffined paper is merely interposed between the ends of two copper wires it still will be a condenser having a definite capacity. There is no need to use paraffined paper to form a condenser. Two metallic plates may be insulated and placed parallel to each other: if they be connected one to one pole of a battery, the other to the other, a current will exist for a finite time; if they are now disconnected from the battery and connected to each other through a suitable instrument for measuring the time integral of current, a return current exists for a finite time. Consider now a condenser consisting of two parallel plates of very large area placed near to each other: its capacity varies as the area of the plates, and it also varies inversely as the distance between them—laws precisely analogous to those of conductors. It also, as might be expected, depends upon the material interposed between the two plates. Since the capacity of the condenser varies as the area, and inversely as the distance, whatever be the substance between the plates, there is a constant for every substance analogous to the specific conductivity of the material; this constant is called the "specific inductive capacity" of the material.

That the magnetic effects of the transient galvanic currents in insulators are the same as of corresponding currents in conductors, is a fundamental law at the basis of Maxwell's theory of electro-magnetism; it was, I believe, first definitely propounded by Maxwell, and it is one over which some of the most eminent minds have since contended. You will notice how, if the subject be arranged in the unhistorical, but not unnatural, way I have adopted this evening, this law drops so easily into its place, that its assumption would be almost unnoticed. Why should we assume that the magnetic effects of electric currents in insulators and in conductors are different? By far the most natural assumption is that they are the same.

Maxwell's electro-magnetic theory of light—one of the great generalisations of the century—is directly based upon this law. What is this theory? It is simple enough in its character, broadly stated. Suppose you have an electrical disturbance, a transient galvanic current, in any dielectric, this disturbance—assuming the magnetic effects of currents in dielectrics to be the same as in conductors—will be propagated as a wave in directions perpendicular to the direction of the currents. The velocity of the wave can be calculated from purely electrical experimental knowledge; it is 3×10^{10}, if the dielectric be a vacuum. The velocity of light is 3×10^{10}. We know nothing of what light is; we do know that it is a wave, and that the disturbance is perpendicular to the direction of transmission of the light; but what is the ultimate nature of that disturbance we have no idea from the standpoint of the science of light. Is it not infinitely probable that the waves of light are none other than the electrical waves which we know must exist, and must be propagated with the observed velocity of light? And, mark, this theory demands no ether; it merely identifies the phenomenon of light with known phenomena of electricity and magnetism; it demands no knowledge of the fundamental nature of these phenomena, it demands only a knowledge of their laws. Surely Maxwell's generalisation is similar both in kind and in magnitude with the discovery of universal gravitation. Newton showed that the force which caused any body—an apple, if you like—to fall on our earth was the same as the force which kept the planets in their courses. He gave no theory of how that force was transmitted, nor of what, in its inmost nature, it was. He merely identified the two forces as one. So now Maxwell's theory may be shorn of all reference to an ether; it identifies as one the forces whereby light is transmitted and the forces with which we are familiar in our dynamo machines, and shows that they are measurably the same. It does not pretend to say what electricity and magnetism are, or what, if any, is the medium whereby the disturbances on which both depend are transmitted. Whether the postulate of an all-pervading ether be, or be not, a metaphysical necessity, surely it is well for the practical man and the physicist to leave the question to the metaphysician. My point, however, is that the hypothesis of two electric fluids and the ethereal theory of light have rather delayed than accelerated the acceptance of a great generalisation. Let us now return to elementary matters.

I have just stated that the capacity of two large plates near and parallel to each other varies as their area, and inversely as their distance. In what units are we to measure capacity? If we have already chosen a unit of resistance, a unit of capacity must follow from that through the units of potential and current. It would be the capacity of that condenser which, if brought to unit difference of potential, would discharge a galvanic current whose time integral was unit current for unit time. It would be quite natural to define as a unit of resistance the resistance between opposite faces of a cube of a defined substance at a defined temperature, in which case the dimensions of resistance would be the reciprocal of a length. Or, again, we might lay by a piece of wire and carefully preserve it, and say, let that be our unit of resistance. Then resistance would have no dimensions in terms of mass, length, and time—it would be a fundamental unit. In either case we should introduce physical constants into our formulæ for electrostatic and electromagnetic phenomena, depending in the one case on the physical properties of the substance chosen, in the other on its actual dimensions. If the history of electrical discovery had followed the course adopted in this discourse, and it had become necessary to fix upon a unit of capacity at the point we have now reached, it would have been very natural to have taken the capacity per square centimetre of two plates at unit distance, the insulating medium being vacuum. We shall see shortly what is the actual electrostatic unit which has been chosen, but the fact is there is nothing specially natural, nothing specially absolute in the choice.

It is a matter of very great interest to know the capacities of condensers of various forms. Let us take, for example, the case of a sphere enclosed concentrically within another sphere. If r_1, r_2, be the radii of the two spheres, the result is that the capacity varies as $\dfrac{r_1 r_2}{r_2 - r_1}$; if r_2 be very great in comparison with r_1, the capacity varies as r_1. The electrostatic unit of capacity ordinarily adopted as absolute is the capacity of a sphere of unit radius enclosed concentrically in a very great sphere, with vacuum as insulator. From this, remembering that the product of a difference of potential and a galvanic current shall be power in ordinary mechanical units, all other units will follow. Thus time integral of current divided by difference of potential or capacity is a line; therefore conductivity is a velocity, and resistance the reciprocal of a velocity. Difference of potential multiplied by current is power, and has dimensions ML^2/T^3, but potential divided by current has dimensions T/L; therefore current squared has dimensions ML^3/T^4, and current $M^{\frac{1}{2}}L^{\frac{3}{2}}/T^2$, and so on.

Again, the capacity of two plates, area A, distance x, can be deduced from that of two concentric spheres at a small distance apart: it is $A/4\pi x$ in these units.

We may now prove that mechanical forces must exist between the plates of a charged condenser. Let a condenser be formed of two parallel plates of area A at a distance x from each other, the distance x being capable of variation; let the condenser be charged to a given difference of potential, and then insulated, and the position of the plates in relation to each other varied. If the condenser be then discharged, it is found, first of all, that the total discharge is the same however the position of the plates be varied. Suppose now that the condenser be charged to a difference of potential E, and a quantity, or time integral, of current Q, the work done in so charging the condenser will be $\frac{1}{2}EQ$. Let now the condenser be insulated, and the distance of the plates increased to $x + \delta x$: the capacity of the condenser will be diminished in the ratio $x/(x + \delta x)$; and therefore, since the quantity remains the same, the difference of potential will have increased to $E(x + \delta x)/x$. Now discharge the condenser. The work done by the discharge will be $\frac{1}{2}EQ(x + \delta x)/x$, or an amount of work $\frac{1}{2}EQ\delta x/x$ greater than was done in charging the condenser. This work must come from somewhere. The only way in which it can be introduced is by the work done in separating the plates from each other. To increase the distance between the two plates by an amount δx therefore requires that we shall do mechanical work $\frac{1}{2}EQ\delta x/x$ upon the plates. This mechanical work is done through a distance δx, therefore a force must be exerted in separating equal to $\frac{1}{2}EQ/x$; hence it follows, as a matter of necessary consequence from the laws of the capacity of conducting surfaces in the presence of each other, that those surfaces shall when charged act mechanically upon each other; and the amount of mechanical force is determined.

Now $\dfrac{EA}{4\pi x}$ is equal to Q, hence the force between the plates is $\dfrac{2\pi Q^2}{A}$. Let us now suppose that matter is laid upon each plate of density $\dfrac{Q}{A}$, and that the matter on one plate attracts that on the other plate with a force varying as the inverse square of the distance: the resultant attractive force between the plates is exactly $\dfrac{2\pi Q^2}{A}$; hence we should accurately express the facts of attraction of two plates if we said there is on each plate a charge of electric substance, and every particle of substance on the one plate attracts every particle on the other with a force varying inversely as the square of the distance. But that the facts can be so expressed is no evidence that any such electric substance exists.

I do not propose to touch on electro-magnetic and magnetic phenomena; I would only point out this: Just as electrostatics has been based on action at a distance between hypothetical electric fluids, so magnetism has been based on the action at a distance of hypothetical magnetic fluids. There can be no doubt that the science could be easily rearranged, and that the phenomena of electro-magnetism could be made fundamental, and that the facts concerning permanent magnets could be made subsidiary thereto, exactly as we have deduced this evening the facts of electric attraction from the phenomena of current-electricity. I believe that in the future this is the way the science must be arranged for practical men. They are but little concerned with the attractions between charged spheres and the like, and very much concerned with the laws of currents and differences of potential and capacities. The theory of permanent magnets has for the modern electrical engineer only a general interest, but the laws of magnetic circuits and induction therein are all-important. It would seem that the subject needs rearranging, and that that which is most important should be put most prominently forward. I even venture to think that this would be no loss to pure science; we should see with increased clearness how little we know what electrical phenomena really are, and how much we know of how electrical laws work. We should see how much there is that is arbitrary and conventional, and by no means absolute, in the system of units we have adopted. The science of electricity is so essential to the practical electrical engineer that it should be arranged so that he can learn in logical order what he most wants to know. In what I have briefly indicated there will be something which is trite to all, and something perhaps not easy to comprehend when stated too shortly. My aim has been rather to excite in your minds discontent with the way in which the science of electricity is usually arranged, in the hope that such discontent may take an articulate form, and become a demand for something more convenient for your particular purpose. When the demand becomes clear it will not be long before it is supplied*.

* I am informed by Professor S. P. Thompson that he has for some time past in his lectures begun the subject of electricity with the consideration of currents. My attention has also been called to Professor Ayrton's excellent book on *Practical Electricity* (1886), and to Dr Walmsley's book on *The Electric Current* (1894), with the contents of which I ought to have been, but was not, acquainted, in which a similar arrangement of the subject has been adopted.—J. H.   *Jan. 22nd.*

16.

PRESIDENTIAL ADDRESS TO THE JUNIOR ENGINEERING SOCIETY, 4TH Nov., 1892, ON THE COST OF ELECTRIC SUPPLY.

[From the *Transactions of the Junior Engineering Society*, Vol. III., Part I., pp. 1—14.]

THE interests of an Engineer are many sided. If he is to successfully use the forces of nature for the service of man he must understand how those forces work; he must in fact be scientific. It may be that his ideas are arranged differently from the ideas of those who study science for its own sake, and without regard to practical applications, but if he is to succeed they must be so arranged that he can deduce from knowledge already acquired, knowledge which is applicable to new cases which have not as yet come under his observation. The Engineer who can only do that which he has seen done before may be a practical man, but he will always belong to a lower grade of his profession. The scientific Engineer is one who by his knowledge of nature is able to deal with new engineering problems and provide useful solutions of those problems. But a practical man must be something more than a man of science, or rather he must look at matters from a different point of view. He cannot choose some feature of a problem, concentrate all his attention upon that, and leave other matters out of consideration, which is the process by which most scientific advance has been made; but he must always deal with the whole matter before him and leave no relevant question out. But an Engineer may be scientific inasmuch as he has knowledge of nature and the power of applying that knowledge in new cases; he may be practical in the sense that the means he devises to attain his ends may be complete at all points, and not break down from trifling defects, and yet may find that there are other subjects which he has to consider. Our complete Engineer must give his attention to commercial matters as well; he must know if, when he has devised the means to attain the ends in view, those ends when attained will result in a profit. He must recognise the conditions which render an undertaking economical to work, and which secure that it shall bring in a large return. When it has been my lot to address Engineers I have usually directed attention to some scientific point which I thought would be of interest to them. This evening I should like to go to the other extreme and deal with a purely commercial question, with a matter into which no science enters, and which relates entirely to pounds, shillings, and pence.

You are all of you familiar with the fact that the expenses of an undertaking may be broadly divided into two classes. On the one hand there are expenses which are quite independent of the extent to which the undertaking is used, and on the other, expenses which are absent unless the undertaking is used and which increase in proportion to the use. For example, the charges for interest on the construction of a bridge are the same whether that bridge is used much or little or at all, and the cost of maintaining the bridge is also practically independent of its user. The same is true in a large measure of a harbour or a dock. Such undertakings lie at one extreme of the scale. It is less easy to find good examples at the present day of the other extreme, as nearly all undertakings with which Engineers have to deal require the employment of some capital, and there will be a fixed charge for the use of that capital and for maintaining against the assaults of time the things in which the capital is embodied. But we can readily see for example in the case of a cotton mill that, if on the one hand there are expenses for interest and dilapidation which are independent of the amount of yarn actually manufactured in a given factory, there are other expenses for material and labour, and even for actual wear of machinery which will be very nearly proportional to the output. Undertakings vary enormously in the proportion of these two classes of expenses, in some the expense is quite independent of the extent of the user, in others it is for the greater part proportional to the user.

But undertakings differ from each other in another respect. In some cases the service which the undertaking is designed to render can be performed at a time selected by the undertakers; in others at a time selected by him to whom the service is rendered. In the case of most manufactures it matters not if the thing made is made to-day or to-morrow, in the morning or the evening, for it will not be used for a month hence perhaps; the thing can in fact be extensively stored and kept till it is wanted. Other services must be rendered at the moment the person served desires. For example, the Metropolitan District Railway must be prepared to bring in its thousands of passengers to the City at the beginning of the day and to take them back in the evening, and for the rest of the day it must be content to be comparatively idle. In this case the services cannot be stored. The line must be of a carrying capacity equal to the greatest demand, and if this be great for a very short time the total return for the day must be small in comparison with the expense of rendering the service. In such a case it would not be inappropriate to charge more for carrying a person in the busy time than in the slack time, for it really costs more to carry him.

Let us see how these considerations apply to the supply of electricity for lighting. Electrical Engineers now realise that they have to provide the same plant and no more to give a steady supply day and night as to give a supply for one hour out of the twenty-four. They also now realise that if they are to be ready to give a supply at any moment, they must burn much coal and pay much wages for however short a time the supply is actually taken. Indeed, the term "load factor" proposed by Mr Crompton is as constantly in the mouths of those who are interested in the supply of electricity, as volt or ampère or horse-power. The importance of the time during which a supply of electricity is used was so strongly impressed on my mind years ago that in 1883 I had introduced into the Provisional Orders with which I had to do, a special method of charge intended to secure some approach to proportionality of charge to cost of supply. Unfortunately the orders of that day all came to nought.

A supply of electricity must be delivered at the very moment when the consumer chooses to use it, and as long as and no longer than he pleases to use it; it cannot be very readily or cheaply stored, and much of the cost of production is the fixed charge for plant and conductors. Furthermore the provisional orders require that the supply shall be available at all hours; hence coal must be consumed and workmen must attend, though but few consumers are drawing a supply. The service of supplying electricity has from an economic point of view a great deal of similarity to the service of providing a breakwater for a harbour. A great deal of the expense is independent of the number of hours in the day during which the supply is used. To put it in another way, the cost of supplying electricity for 1,000 lamps for ten hours is very much less than ten times the cost of supplying the same 1,000 lamps for one hour, particularly if it is incumbent on the undertaker to be ready with a supply at any moment that it is required.

The actual importance of considerations of this kind can only be realised by examining figures. The figures may as well be estimated figures, because the circumstances vary from one neighbourhood to another. No criticism of the details of the figures will affect the general character of the conclusion. Let us then imagine a station capable of supplying 40,000 sixteen-candle lamps at one time, with mains and spare machinery enough to ensure that the supply shall not fail, and let us see what the charge for running such a station will be; firstly on the hypothesis that it is always to be ready to supply the 40,000 lights at half-an-hour's notice day or night but that the lights are hardly ever actually required; secondly on the hypothesis that the 40,000 lights are steadily and continuously supplied day and night. These are the two extreme cases possible. In the former, the load factor is nil; in the latter it is 100 per cent. If the charge is by meter at 8d. per unit in the former case, the revenue will be nil; in the latter it will be £730,000 a year.

We are going to divide the cost of supplying electricity into two parts; a part which is independent of the hours the supply is used, and a part which is directly proportional thereto; and we are going to estimate the amount of each element. It is for the purpose of ascertaining these elements that we consider two quite hypothetical cases; cases which can themselves never actually occur.

We must first have an idea of the capital outlay required. To provide the maximum of 40,000 lamps we need to deliver 2,500 units per hour, and we may estimate the capital outlay as follows:—

	£
Land	25,000
Buildings	15,000
Boilers and Pipes	14,000
Engines	24,000
Dynamos	15,000
Switchboard and Instruments	2,000
Feeders and Mains	50,000
	£145,000

Let us deal with the annual charge for each item of capital separately on the two hypotheses. The charge for land and for buildings including repairs is clearly the same in the two cases, say at 4 per cent. £1,000 for the land, and at 10 per cent. £1,500 for the buildings. The boilers, engines, and dynamos will have a charge for interest, and a charge for writing off or amortization as the French call it, that is, for writing off the value of the plant before the time at which it becomes antiquated,—exactly the same in the two cases. The boilers too will require exactly the same repairs whether they are merely keeping steam or whether they are generating steam continuously; but the machinery will certainly require more for repairs and renewals if it is all running than if a part only is running without load and the rest is standing ready for a load if required. I take 4 per cent. as the charge for interest; 3 per cent. for amortization; 8 per cent. for repairs and maintenance. Of the repairs of engines and dynamos I assume that 2 per cent. will be applicable if the plant runs light, the remaining 6 per cent., if it is fully and continuously loaded. The expenses connected with conductors and switchboard, etc., will be exactly the same whether the current is passing or not; these I take at 15 per cent. The rates I put down at £500 a year. The account then for the fixed charges already enumerated would stand as follows :—

	Running Light. £	Fully Loaded. £
Land	1,000	1,000
Buildings	1,500	1,500
Rates	500	500
Boilers	2,100	2,100
Switchboard and Conductors	7,800	7,800
Engines	2,160	3,600
Dynamos	1,350	2,250
	£16,410	£18,750

We now come to a most important item in the account, the coal. There is no doubt that with uniform and continuous load a unit of electric energy—$1\frac{1}{3}$ horse-power for one hour—can be produced for less than 3 lbs. of coal; it is also pretty much admitted that with a load factor of about 12 per cent., but continuous maintenance of pressure, the consumption of coal in good practice is something like 7 lbs. That is to say, to keep the boilers warm, turn round the machinery for 24 hours, and deliver full current for 24 hours, will require 72 lbs. of coal per kilowatt; whereas to keep the boilers warm, turn round the machinery, and deliver current for 3 hours, will require 21 lbs. of coal. The boilers being kept warm, it will take 51 lbs. of coal to generate steam enough to give a unit per hour for 21 hours; 58 lbs. to give a unit per hour for 24 hours; subtracting this from 72 lbs., the amount required both to generate steam and keep the boiler warm, we may infer that to keep the boiler warm and merely turn the machinery in readiness to meet a demand will take about 14 lbs. of coal per day for every unit per hour the plant is capable of producing. In 1889, for the Society of Arts, tests were made of a Paxman compound engine, from which it appears that a boiler which when fully loaded consumed 40 lbs. of coal per hour, required 4 lbs. per hour to keep steam up to normal pressure when the engine was standing: that is, 10 per cent. of the coal used was used to maintain the steam pressure. Remembering that in addition we keep some of our machinery moving, this may be said to confirm the figures adopted. Thus if the plant runs light all the year round 12,775,000 lbs., or let us say 6,000 tons of coal will be consumed. If the plant runs fully loaded 65,700,000 lbs., or let us say 30,000 tons would be consumed. If we suppose the coal to be best smokeless it might cost 20s. per ton. Next we have water, oil and petty stores; say £600 and £3,000 in the two cases. Wages will be a little less if we run light than if we run fully loaded, and of course will largely depend on local circumstances; let us say £5,000 and £7,500 in the two cases. This gives us substantially all the expenses which have to be met and our account will then stand thus:—

	Running Light. £	Fully Loaded. £
Fixed Charges............	16,410	18,750
Coal	6,000	30,000
Stores	600	3,000
Wages	5,000	7,500
	£28,010	£59,250

Thus the cost of merely being ready to supply 2,500 units per hour at any moment throughout the year will be £28,010, and the cost of actually supplying 2,500 units per hour for every minute in the year will be £59,250. The undertaker therefore who incurs the liability to supply, ought to receive £11 per annum per unit per hour from those on whose behalf he incurs the liability, and if he receives the £11 he need not charge more than $\frac{1}{3}d.$ per unit for what he actually supplies, to cover his expenses. That these figures are fair approximations can be seen as follows: according to this calculation the cost of supplying 2,500 units for one hour per day is £28,010 + 2,500 × 365 × $\frac{1}{3}d.$ = £29,277, and the charge for the service at 8d. a unit would be £30,417; it is doubtful if such a supply would pay. On the other hand an indicated horse-power on such a scale could certainly be supplied continuously for from £12 to £14 per annum, and according to this calculation an electrical horse-power will cost just under £18 per annum. No account is taken of expenses peculiar to companies, such as directors' fees and the cost of forming the company. It will also be noted that it is assumed that accumulators are not used.

The charge for a service rendered should bear some relation to the cost of rendering it. If it is a matter of open competition the matter will settle itself, for no one will for long be able to supply some customers at a loss, and recoup himself by exorbitant profits from others. If the matter be a case more or less of monopoly, the adjustment is less certain; thus the Post Office charges $\frac{1}{2}d$. postage for a printed circular and $1d$. for a written letter, the two costing the Post Office exactly the same. What a boon to the public it would be if the Post Office would charge more for printed trade circulars, which in nine cases out of ten are a nuisance to those who receive them. The supply of electricity is not quite a monopoly; companies compete with each other, and there is always the competition with other methods of illumination such as gas and paraffin. It is clearly to the advantage of the undertaker to secure all those customers whom it pays best to supply, and as far as may be, to compel those who are unremunerative to adopt these other methods. The ideal method of charge then is a fixed charge per quarter proportioned to the greatest rate of supply the consumer will ever take, and a charge by meter for the actual consumption. Such a method I urged in 1883, and obtained the introduction into certain Provisional Orders of a clause sanctioning "a charge which is calculated partly by the quantity of energy contained in the supply and partly by a yearly or other rental depending upon the maximum strength of the current required to be supplied." In fixing the rates of fixed charge it must not be forgotten that it is improbable that all consumers will demand the maximum supply at the same moment and consequently the fixed charge named might be reduced or some profit be obtained from it. There is no object in reducing the cost of electricity for lighting in the case of any customer much below the cost of equivalent lighting by gas, unless there are competitors in the field willing to do it, hence the current charge proportioned to the power supplied may safely be increased. In certain recent cases in which I am acting as engineer, the Board of Trade have sanctioned on my application, "for each unit per hour in the maximum power demanded, a charge not exceeding £3 per quarter, and in addition for each unit supplied, a charge not exceeding two pence." It is sometimes said as an objection to this method of charge, the public will object to pay a fixed charge whether they make use of their lights or not, and that in fact they will not pay it. The best answer that can be made is to give everyone the choice of being charged the maximum simple rate provided by the Order, or by the compound rate, as they prefer. What is wanted is not so much an increased charge for those consumers whose lights are used for a short time, as such a special reduced charge for those whose lights are used long as will induce them to use the supply.

It is instructive to compare the cost to different classes of consumers of electricity and gas for lighting with 16-candle gas. Flat flame burners must be large and of first-rate quality to give more than two candles per cubic foot of gas per hour; the large majority of burners give much less than this even at their best, and as a rule the pressure of the gas is not regulated and much gas is wasted as far as the production of light is concerned. Incandescent lamps give about one-quarter of a candle per watt; hence a Board of Trade unit is equivalent to 125 cubic feet of gas. Thus we readily arrive at the following comparative table, the charge being at the rates recently sanctioned by the Board of Trade:—

Hours of use per annum.	Load Factor.	Price of Gas at which cost of lighting by electricity and 16-candle gas are equal.
480	5·5	5s. 4d.
960	10·9	3s. 4d.
1,440	16·4	2s. 8d.
1,920	21·9	2s. 4d.
2,880	32·9	2s. 0d.
3,840	43·8	1s. 10d.
7,680	87·6	1s. 7d.

In the accompanying curves are shown the cost of production, and the charge per unit at the compound and simple rate. The ordinates represent pence and the abscissæ the number of hours per annum the supply is used.

It is obvious that those whose user is long will find the electric light economical to themselves and that it will be profitable to the undertaker. With a cheap light which is free from the products of combustion there will be extensions for the hours of use. Shops may find it worth while to continue the light after closing, as an advertisement.

We have so far assumed that the supply of electricity is carried on without the aid of accumulators. Let us first compare the cost of an electric accumulator with the cost of a gas-holder containing the same possibility of producing light. A gas-holder is at present being put up in Manchester to hold 7,000,000 cubic feet of gas and is to cost complete with its tank £60,000. With 16-candle gas seven million cubic feet are equivalent to 56,000 Board of Trade units. Accumulators, capable of storing a ten hours' supply, cost about £50 per unit. The equivalent accumulator will therefore cost about £280,000. But this is not all; the gas-holder is comparatively permanent; the accumulators require

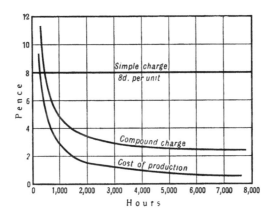

frequent renewals and repairs; the gas-holder gives back all the energy put into it: the accumulators waste at least 20 per cent.; the gas-holder may be emptied as fast as you please; the accumulators, not faster than a certain rate without diminishing their capacity. Taking all into consideration, the cost of storing energy by the aid of accumulators and storing it in a gas-holder are quantities of a different order of magnitude. If no gas-holders were used, and all the gas had to be made just as it was wanted, its cost for lighting would be several fold what it now is, even if gas-producers could be found capable of instantly varying the supply as the demand varies. The gas-producing plant would have to be enormously increased; so would the size of the mains, and so would the wages of labour. If electric power could be stored as cheaply as gas, there would soon be little hope that the gas companies would maintain their dividends.

Let us see from a financial point of view whether accumulators can be used economically for storing up electrical power continuously produced during the 24 hours, and used rapidly for a short time.

Assume that the whole of the plant with the accumulators is capable of supplying 40,000 lights for ten hours continuously, and that during that time half the power is supplied from the accumulators. Ten hours in the twenty-four hours is not an unreasonable allowance, for we have melancholy experience in London of continuous fog for days, and this would tax the plant we are considering to the utmost. We are to be ready then at any time on short notice to supply 40,000 lights, and to continue to supply them for 10 hours. Compare the cost firstly of maintaining this state of readiness with the accumulators and with a plant without accumulators. We shall require a battery capable of giving 1,250 units for ten hours; such a battery costs not less than £50 per unit, or in all £62,500. To maintain it, will cost from 10 to 15 per cent. on the cost; there will also be interest on the outlay and amortization, say in all 20 per cent. or £12,500 a year. If we assume that the batteries are distributed at the various points of the system of conductors, we may also assume that the charges for land and buildings will be much the same as for the plant without accumulators. The boilers, engines, and dynamos will be just one-half. The switchboard and instruments will be much the same. But the conductors will be reduced, smaller or shorter feeders being necessary, probably £40,000 will go as far with accumulators as £50,000 without. The coal bill may be dispensed with entirely, as we may assume that steam could always be got up during the time in which the demand increased from nothing to one-half of the maximum, and that therefore all the coal burned can be assumed to be burned for producing current. That is to say, we assume the quantity of coal burned is proportional to the quantity of electric energy, and that therefore when no electricity is actually used, no coal will be burned. The wages may be reduced, for we have only to be ready to run half the plant, and a small wage will suffice for attendance on the accumulators. The wages of linesmen and the like will remain the same. Assume the total wages to be £3,500 instead of £5,000. The account will then stand thus:—

Land	1,000
Buildings	1,500
Rates	500
Accumulators	12,500
Boilers	1,050
Engines	1,080
Dynamos	675
Switchboard	300
Conductors	6,000
Wages	3,500
	£28,105

practically the same result as we obtained before.

Now consider another hypothetical case, which of course can never occur in practice. We are to supply 40,000 lamps for ten hours every day with the plant just described, charging the accumulators during twelve and a half of the fourteen hours during which the light is not required, twelve and a half hours' charging giving ten hours' discharge of the same energy. The coal would cost the half of £30,000 if the machinery had to run the whole of the 24 hours. It has to run $22\frac{1}{2}$ hours, but the boilers have to be kept warm the whole time, hence the coal will cost the half of £6,000 for keeping the boilers warm, and $\frac{22\frac{1}{2}}{24}$ of the half of £24,000 for generating steam. The wages may fairly be taken as £4,750, and the account will stand:—

Land	1,000
Buildings	1,500
Rates	500
Accumulators	12,500
Boilers	1,050
Engines	1,800
Dynamos	1,125
Switchboard	300
Conductors	6,000
Wages	4,750
Coal	14,250
Stores	1,425
	£46,200

The cost of supply for the same ten hours without accumulators would be as follows:—

	£
Land	1,000
Buildings	1,500
Rates	500
Boilers	2,100
Switchboard and Conductors	7,800
Engines	2,760
Dynamos	1,725
Coal	16,000
Stores	1,600
Wages	6,000
	£40,985

a cost of about 11 per cent. less than where accumulators are used.

Putting it another way, the cost of being ready to supply and to continue to supply, is about the same whether accumulators are used or not; the additional cost of actually supplying current is about 40 per cent. more where accumulators are used than where they are not used. It may be safely inferred that the use of accumulators does not seriously alter the conclusions I have drawn as to the proper method of charging consumers for a supply of electricity.

The question of whether the great cost of a supply for short hours can be removed by the use of accumulators may be looked at in another way. Will it pay a consumer to put in his own accumulators and charge them from the station supply? We may reasonably suppose the undertaker will remit the fixed charge in consideration of the consumer only taking his current at slack times. His accumulators if they are to be of capacity to maintain his supply through a foggy day will cost him £50 per unit per hour (or per kilowatt) and the annual charge in respect of them will be £10 per year, to which if we add a rent for the space the battery occupies, gives us a charge not differing materially from the fixed charge made or suitable to be made by the undertaker. But in order to obtain 2d. worth of electricity he must purchase 2$\frac{1}{2}d$. worth for charging his battery.

A word or two more about the use of accumulators. These have certainly improved, and they will continue to improve. They will become more durable and more economical of power in working, and their first cost will become less. An inspection of my tables of cost shows that a very little improvement would render them valuable even in very large stations for the mere purpose of diminishing the machinery required, by storing the energy developed at slack times to be used in busy times. The certainty of improvements in accumulators, and the possibility that the improvement may be considerable, is a strong argument for the use of the direct current wherever it is not precluded by the distance of transmission being too great.

It will be noted that I have assumed a very large station. Accumulators have another use which greatly increases their advantage in smaller stations. There are many hours in the twenty-four when it is absolutely certain that the demand will be small. If accumulators are used, the attendance of the staff may be dispensed with during those hours, and a considerable sum in wages will be saved. The proportion of wages to the whole of the charges is much greater in small stations than in large. In most small stations giving continuous supply, accumulators ought to be used notwithstanding their expenses and defects, and I believe the day is not far distant when they ought to be used in connection with most large stations also.

If instead of a continuous current, an alternating current with transformers is used, the modification in the account will be that the cost of conductors will be diminished, but the cost of transformers will have to be added. If the distances are small, the increased cost of transformers will exceed the saving in the conductors; if the distances are considerable, the cost of transformers will be less than the saving of conductors. In both cases the general character of the result will be the same as before, the cost of being prepared to give a supply will be considerable, and the cost of actually giving the supply will be much smaller than is generally supposed. Indeed with the alternating current this peculiarity will be even more marked, for the machinery has not only to be kept in motion however small the consumption may be, but a certain current must be maintained in every transformer. With the best transformers, this current may only have an energy

1½ per cent. of the energy of the current when the transformer is fully loaded. This would increase the coal bill in the case considered by about £500 per year whether the supply was used or not.

It is possible, indeed probable, that some of my assumed figures may be shown to be too high or too low for the generality of cases. It is of no moment; let each one take any figures he pleases within reason; let him assume that the supply of electricity is made by any system he pleases; he will arrive at a result broadly similar to mine. To be ready to supply a customer with electricity at any moment he wants it will cost those giving the supply not much less than £11 per annum for every kilowatt, that is for every unit per hour, which the customer can take, if he wishes, and afterwards to actually give the supply, will not cost very much more than ½d. per unit. This is the point I have been labouring to impress, for I take it, it is essential to the commercial success of Electric Supply. It is hopeless for electricity to compete with gas in this country all along the line, if price is the only consideration. But with selected customers, electricity is cheaper than gas. Surely it is the interest of those who supply electricity to secure such customers by charging them a rate having some sort of relation to the cost of supplying them.

17.

THE RELATION OF MATHEMATICS TO ENGINEERING.

[From the *Proceedings of the Institution of Civil Engineers*, Vol. CXVIII. Session 1893—94. Part IV.]

MATHEMATICS has been described in this room as a good servant but a bad master. It will be my duty this evening to prove by suitable illustration the first half of the proposition, and to show the service mathematics has rendered and can render to engineers and engineering.

In our Charter the Institution of Civil Engineers is defined as "A society for the general advancement of Mechanical Science, and more particularly for promoting the acquisition of that species of knowledge which constitutes the profession of a Civil Engineer, being the art of directing the great sources of power in Nature for the use and convenience of man, as the means of production and of traffic in states both for external and internal trade, as applied in the construction of roads, bridges, aqueducts, canals, river navigation and docks, for internal intercourse and exchange, and in the construction of ports, harbours, moles, breakwaters and lighthouses, and in the art of navigation by artificial power for the purposes of commerce, and in the construction and adaptation of machinery, and in the drainage of cities and towns." No better definition can, I think, be found for our profession than that it is the art of directing the great sources of power in Nature for the use and convenience of man. It covers all that the widest view of our work can include, and it excludes those applied sciences, such as medicine, which deal with organized beings. Mathematics has to deal with all questions into which measurement of relative magnitude enters, with all questions of position in space, and of accurate determination of shape. Engineering is a mathematical science in a peculiar sense. Medicine, the other great profession of applied science, has but little to do with questions of measurement of magnitudes, or of geometry; but the engineer finds them enter into everything with which he has to deal, and enter in the most diverse ways. The thing he has to determine is that the means he employs is enough and not unnecessarily more than enough to attain the end in view. For this he must numerically measure the end and the means and see that they are justly proportioned to each other. It is useless this evening to waste time proving, what all will admit, that no one can be even the humblest engineer without a knowledge of arithmetic and enough of geometry to enable him to read a drawing, that some trigonometry, some rational mechanics and a knowledge of projections, is a very useful part of the mental equipment of a draughtsman. It is hardly necessary to call attention to the great economy in the labour of calculations effected by the use of logarithms, a mathematical instrument for which we are indebted to Napier. We may with more profit examine what use the higher mathematics can be to the practical engineer, and what has been done in the past for engineering by its aid.

Judging from etymology, mathematics must have been begun by engineers; for surely geometry is the work of the earth measurer or land surveyor. But since the prehistoric times when geometry was initiated, engineers have not added much that is new to mathematics. They have rather sought amidst the stores of the mathematician and selected the handiest mathematical tool they could find for the particular purpose of the moment, but have done little or nothing in return in the way of improving the tools which they borrow. In this respect the relation of engineering to mathematics differs much from its relation to experimental physics. In electricity, magnetism and heat, engineers have from their practical experience repeatedly corrected the ideas of the theorists, and have started the science on more accurate lines. If our subject to-night had been the use of the practical applied science of engineering in promoting the development of pure mathematics, we should speedily find that there was hardly any material for discussion. The account being all on one side, let us see to what the debt of the engineer to the mathematician amounts.

There is no department of practical engineering in which the application of mathematics is more familiar than in that which relates to the calculation of the strength and rigidity of structures of various kinds. It is impossible to take up any book dealing with the subject without finding that it is crammed either with mathematical formulæ, or with geometrical figures. The question is not whether mathematics is necessary to an adequate comprehension of the subject, but whether analytical or purely geometrical methods are more convenient. Of course one might occupy many lectures in discussing the practical application of mathematics to the question of bridge building, roofs, guns, shafting, and the like. Our object must be to illustrate by various examples rather than to attempt anything like a complete discussion.

Consider the case of a long strut, so long that its transverse dimensions can be regarded as insignificant in comparison with its length. Whilst the strut remains perfectly symmetrical about its middle line, its strength will depend only upon the resistance of the material to crushing. Everyone knows that this would be an inadequate conclusion; we have to consider another element, namely, its stability, that is, we must examine what will happen to the strut if from any cause it is displaced somewhat from the direct line between its extremities. A mathematical discussion of the question results in a differential equation of the second order with one independent variable. Upon consideration of this we are enabled to see that if the thrust upon the strut be less than a certain critical value, a slightly bent strut will tend to return to its straight condition; but that if the thrust upon the strut be greater than this critical value the displacement will tend to increase, and the strut will give way. Further, that the critical value will depend upon whether one or both of the two extremities of the strut are held free, or whether they are rigidly attached by flanges or otherwise, so that the direction of the axis of the strut at this point must remain unaltered. Again, we infer that if the ends are held rigidly fixed, the length of the strut may be twice as great for a given critical value of the thrust as if the two ends are free to turn. We can also infer what the critical value will be for struts of various lengths and of varying cross sections. This critical value depends not upon the resistance of the material to crushing, but upon its rigidity.

Another example, having a certain degree of similarity with the case of struts, is that of a shaft running at a high number of revolutions per minute, and with a substantial distance between its bearings; for simplicity, we will suppose that there are no additional weights such as pulleys upon the shaft. How will the shaft behave itself in regard to centrifugal force as the speed increases? In this case, so long as the shaft remains absolutely straight it will not tend to be in any way affected by the centrifugal force, but suppose the shaft becomes slightly bent, it is obvious to anyone that if the speed be enormously high this bending will increase, and go on increasing until the shaft breaks. In this case also we may use mathematical treatment; we find that the condition of the shaft is expressed by a differential equation of the fourth order, and from consideration of the solution of this equation we can say that if the speed of any particular shaft be less than a certain critical speed, the shaft will tend to straighten itself if it be momentarily bent, but that, on the other hand, if the speed exceeds this critical value, the bending will tend to increase with the probable destruction of the shaft. I do not know that either of these two questions can be properly understood without some knowledge of differential equations.

A problem having a certain analogy to those to which I have just referred is that of hollow cylinders under compression from without, such as boiler tubes. Whether the tubes be thick or thin, so long as they are perfect circular cylinders, they should stand until the material was crushed. But if the tubes are thin, what will happen if the tube from any cause deviate ever so little from the cylindrical form? The solution cannot be obtained without a substantial quantity of mathematics.

The next illustration shows how a mathematical conclusion, correct within the limits to which it applies, may mislead if applied beyond those limits, and how a more thorough mathematical discussion will give a correct result. Considering a case of shafting in torsion it was shown by Coulomb that the stiffness and strength of a shaft having the form of a complete circular cylinder could be readily calculated if the transverse elasticity of the material and its resistance to shearing were known. From the complete symmetry about the axis it is evident that points which lie in a plane perpendicular to the axis before twisting will still be in that plane when the shaft is twisted; it is also clear that the angle through which all points in the same plane move will be the same; hence the problem was as simple as problem could be. But many who had occasion to make use of Coulomb's results gave them an application which was wholly unwarranted. They assumed that they were equally applicable to other cases than complete circular cylinders; they assumed in fact that every point of the material which lay in a plane perpendicular to the axis would remain in that plane when the shaft was twisted, whether the shaft was symmetrical about its axis or not, and they consequently arrived at very erroneous results. That the assumption was erroneous is obvious enough from a consideration of an extreme case. In Fig. 1 is shown in cross-section a hollow cylindrical shaft, which is not complete, but divided by a plane passing

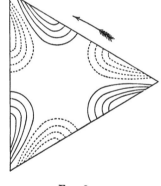

Fig. 1.   Fig. 2.

through its axis. In this case the shaft when twisted will be as illustrated in the side elevation; two points, A and B, were in one plane perpendicular to the axis when the shaft was free from twist; they cease to be in one plane when the shaft is twisted. St Venant* in 1855 investigated the question of shafts without making incorrect assumptions; he expressed the condition of the

* *Mémoires des Savants Étrangers*, 1855; and Thomson and Tait, *Treatise on Natural Philosophy*.

H. I.   18

material by a partial differential equation of the second order, and gave suitable surface conditions. A general solution of the problem for all forms of shafts has not been obtained, but St Venant gives a number of solutions for particular forms, and he obtains some general results of interest. In all cases the stiffness of the shaft is less than would be inferred from an erroneous application of Coulomb's theory. Fig. 2 shows diagrammatically the strain in a shaft of triangular section; the full lines indicate that the parts of the shaft which lay in one plane before twisting when twisted rise above the plane; the dotted lines indicate that they lie behind the plane of the paper. The shearing stress is least at the angles of the triangle, and is greatest at the middles of the sides. At this point then the shaft will begin to break under torsion. The fact is probably well known to men of practical experience, but it is directly contradictory to the conclusion at which one would arrive by a careless use of Coulomb's theory beyond the narrow limits within which it is applicable. The longitudinal ribs which one often sees on old cast-iron shafts are useful enough to give stiffness to the shafts against bending, but are good for very little if torsional stiffness or torsional strength is desired.

Another application of mathematical theory which has been carried somewhat further than the premises warrant is found in the case of girders. It is almost invariably customary to treat a girder as though the sections retain when the girder is bent the form and size which they had before bending. Making this assumption, it is very easy to calculate the strength and stiffness of a girder of any section. Unfortunately, the assumption is untrue; but, fortunately, it is approximately true in the case of most girders with which engineers in practice have to deal. That it is untrue can be readily seen from consideration of a girder of exaggerated form, the section of which is shown in Fig. 3. Any practical man would at once see that the outer parts of the flanges would add little to the strength of the girder, but according to the usual mathematical theory the outer parts of this flange should be as useful as the parts which are nearer to the web. This problem St Venant also deals with in a rigid mathematical manner. Amongst other things, he showed that a girder of rectangular section, such as shown in Fig. 4, would, when bent, take the form shown by the curved lines in the same figure. The last two examples show how a little knowledge may be a dangerous thing, and how easy it is for anyone who attempts to apply mathematics without adequate mathematical knowledge to be misled.

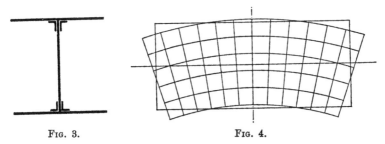

FIG. 3.   FIG. 4.

The theory of thick cylinders under bursting stress from within has many important practical applications to hydraulic presses and to guns. It has been discussed more than once in this room. As usual in considering these cases we are immediately led to differential equations, which here are fortunately solved without serious difficulty, and the solution tells us the whole story. We learn that doubling the thickness by no means doubles the strength of the cylinder. And as a converse, that doubling the strength of the material will permit the thickness to be diminished to much less than one-half. Twenty-five years ago hydraulic presses were mostly made of cast-iron. Many people were not a little astonished at the great reduction in thickness and weight which became possible when steel was substituted for the weaker material. In the case of guns it is well known that greater strength can be obtained if the outer hoops are shrunk on to the inner ones[*]. Mathematical theory tells us what amount of shrinkage should give the best results. It may possibly not be worth while to follow the results of theory precisely, but without the guidance of theory it would not be unnatural to give so great a shrinkage that the gun would be weaker than if no shrinkage were used.

The rolling of ships in a sea-way gives an illustration of a principle which has very varied application in many branches of physics. Suppose a body is capable of oscillating in a certain periodic time, and that it is submitted to a disturbing force of

[*] Lamé, Leçons sur la Théorie Mathématique de l'Élasticité des corps solides. 14th Leçon.

given period, the equation of motion easily shows that the resulting disturbance will be great if the two periods are equal or nearly equal. We meet with the principle in acoustics as resonance. If two tuning-forks are tuned to the same pitch, and one is sounded in the neighbourhood of the other, that other will presently be thrown into vibration by the waves transmitted through the air from the first. You may try a similar experiment at any time on any piano. Strike the higher G in the treble, the sound ceases on raising the finger. Now hold down the middle C, and again strike G, the C string at once takes up the note sounded, and can be heard after the exciting string has been silenced by damping. The same fundamental idea is found in the lunar theory in the term in the equation known as the evection, and again in the theory of Jupiter's satellites. The reason why the metals present in the solar atmosphere give black lines in the spectrum by absorption, corresponding in position with the bright lines in the spectrum which the same metals give when incandescent, is again the same. Gas will absorb or take in from the ether waves of the exact period which it is capable of giving to the ether. The general explanation of all these phenomena is easy. Imagine a pendulum, and suppose it experiences a periodic disturbing force, the first impulse of the disturbing force gives the pendulum a slight swing, the effect of the second impulse depends entirely on when it occurs, it may occur so as to neutralize the effect of the first, or it may occur so as to increase it. If the period of the force is the same as the natural period of the pendulum, the effect of the second, third, and later impulses will be added to the effect of the first, and the final disturbance will be great even though the individual impulses be minute. But the mathematical theory tells us much more than any general explanation can do. It tells us exactly what the character of the effect will be, and its amount if the periods are nearly but not exactly the same. It tells us, too, exactly how friction affects the results. And the beauty of it is that the mathematical theory is much the same in all cases, so that having learned to deal with one case we are enlightened as to a host of others. The oscillating body may be an ironclad, or it may be an atom of hydrogen, the disturbing periodic force may be the waves of the Atlantic, or it may be the waves in the ether occurring five hundred millions of millions of times

THE RELATION OF MATHEMATICS TO ENGINEERING. 277

in a second; it is all one to the mathematician, the treatment is substantially the same.

The question of the speed of ships and the power to propel them is probably more effectually treated by experiment on models, as was done by the late Mr Froude, than by mathematics alone; but in order to learn from the experiments all they are capable of teaching a mathematical understanding is needed. Given that we know by experiment all about a given model, that we know what force is needed to propel it at every speed, we want to know from these experiments how a great ship, 100 times as big but similar in form in every respect, will behave, and here mathematics come in to aid us in making the inference.

The construction of ships at once leads us on to the methods of navigating them. In navigation I should find much material for my purpose, but navigation is not usually included in engineering, but many of the implements of navigation undoubtedly are. The mariners' compass has for ages been the mainstay of the navigator, and a simple enough instrument it was till it was disturbed by the iron of which ships came to be built. The disturbance of the compass by the iron of the ship was first seriously attacked by two senior wranglers, Sir G. Airy and Mr Archibald Smith. The disturbance may be divided into two parts, the first due to the permanent magnetism of the ship, the second to the temporary magnetism induced by the earth's inductive action on the iron of the ship—the first causes the semicircular, the second the quadrantal, error. One has only to open the "Admiralty Manual of Deviations of the Compass" to see how the mathematics of Archibald Smith have accomplished a proper understanding of the subject. The errors of the compass are dealt with in two ways: they are compensated by soft iron correctors, and by permanent magnets so placed as to have an effect equal and opposite to the effect of the temporary and permanent magnetism of the ship. Or they are dealt with by formulæ of correction which enable the error to be calculated when the course of the ship and the conditions of the earth's magnetism are given, or a combination of the two methods is used. Either method is based on Archibald Smith's theory. It is not possible to leave the subject of the mariners' compass without referring to the great improvements of Lord Kelvin. The improvements relate to every part of the instrument, and I venture to say that none of them could have been made by anyone but a mathematician. In order to get his card steady he knew that its period must be different to any possible period of the waves, or he would have the resonance to which I have just referred coming in, so he gave his card a considerable moment of inertia; but this was managed with a light card so that small needles could be used. If the needles are small the correction by soft iron masses and by permanent magnets is easier and more accurate. Then the bowl of the compass had to be suitably carried so that it would not be unduly disturbed by shock, and provision had to be made for damping by fluid friction the oscillations of the bowl if they occurred. Lastly, a most beautiful method of correcting the compass, without taking a sight, was discovered. In every detailed improvement one can detect that the inventing mind was that of a most able and trained mathematician.

An essential of safe navigation is an efficient system of lighthouses. The optical problem of the lighthouse engineer is to construct apparatus which shall usefully direct all the light produced. The present forms of apparatus are in their leading features due to Fresnel, the able mathematician, who established on an absolutely firm foundation the undulatory theory of light. To properly design an optical apparatus formulæ must be used, and the advantage is great if the designer can with ease manufacture the formulæ he requires.

Submarine telegraphy yields some interesting examples of the application of the higher mathematics. When a cable across the Atlantic was first seriously entertained, the first point to be settled was, how many words a minute could be sent through such a cable. This was the most practical question possible. Upon the answer depended the prospect of the cable paying commercially if successfully laid. The matter was dealt with by Professor Thomson[*], of Glasgow, now Lord Kelvin. He showed that the propagation of an electric disturbance in a cable could be expressed by a partial differential equation, and that the solution of this equation under certain conditions applicable to practice could be expressed either by a definite integral or by an infinite series. The values of these were calculated, and hence before an Atlantic cable

---

[*] *Mathematical and Physical Papers*, vol. II. p. 61. Sir W. Thomson.

THE RELATION OF MATHEMATICS TO ENGINEERING. 279

was laid at all it was known how long it would take a signal to reach the opposite shore, and how much its intensity would be diminished in transmission. Referring to Fig. 5, abscissæ repre-

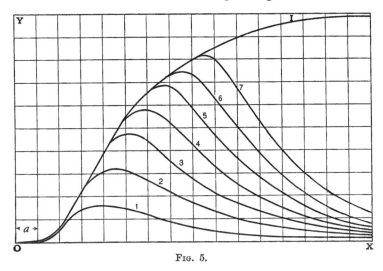

FIG. 5.

sent time, reckoned from the time of making contact at the sending end of the cable, ordinates the currents at the receiving end, curve (I) gives these currents when the contact at the receiving end, after being made, is continuously maintained. It will be observed that for a time a there is hardly any current at the receiving end, and then the current rapidly increases, and attains to half its final value after a time equal to about $5a$. Curves (1)......(7) show the currents at the receiving ends when the contact is made at the sending end maintained for times a, $2a$ ......$7a$ respectively, and then broken. Looking at curve (I) one sees how small is the amount of current and how long it lasts compared with the time during which contact is made. The time a depends on the length and character of the cable; it is equal to $kcl^2 \log_e \frac{4}{3}/\pi^2$, where k is the resistance per unit length, c the capacity per unit length, and l the length of the cable. The knowledge of what is the commercial value of a cable depends on a knowledge of the value of a, and this cannot be obtained without knowing the differential equation $ck \dfrac{dv}{dt} = \dfrac{d^2v}{dx^2}$*, to which I have

* v is the potential, t the time, and x the distance from the sending end of the cable.

referred, and its by no means simple solution either as a definite integral or as an infinite series. So far as I know, this piece of higher mathematics cannot be evaded by any mere elementary treatment. The transmission of disturbance in a cable is quite different from the transmission of sound waves in air, which move with constant velocity. If the cable be doubled in length, it takes four times as long for the signal to pass through it instead of just twice as long, as would be the case if it were a proper wave motion. In fact the time of passage between the making of contact at the sending end of the cable and the beginning of the resulting disturbance at the receiving end, varies as the square of the length of the cable. The mathematical theory is exactly the same as that of the transmission of heat in a plate, one surface of which is suddenly exposed to a temperature different to the temperature of the plate. This is constantly occurring in the application of mathematics—one piece of mathematical work serves for many physical problems having apparently little in common. Fourier long ago discussed the heat problem, little dreaming that his analysis would be just what was wanted for ascertaining how fast signals could be sent across the Atlantic by a system of telegraphy which in his days had not even been projected in its simplest form. The same differential equation also gives the theory of the transmission of telephonic messages through cables; but the solution is then easier, and tells us exactly why it is so much more difficult to speak through 100 miles of cable than through 1,000 miles of overhead line. As I have just stated, the differential equation of the disturbance in the cable is $ck \dfrac{dv}{dt} = \dfrac{d^2v}{dx^2}$. A musical note of period T spoken into the cable through a telephone is properly represented by $A \sin \dfrac{2\pi t}{T}$; the disturbance in the cable will be—

$$v = A e^{-x\sqrt{kc\pi/T}} \sin\left(\dfrac{2\pi t}{T} - x\sqrt{kc\pi/T}\right),$$

as may be easily verified by differentiating. This equation tells us everything. It tells us the rate at which the waves diminish with the distance. This rate increases with the resistance, with the capacity, and with the frequency. If the capacity is at all considerable the diminution is rapid. The velocity of the waves is not the same for all frequencies, as is the case with waves in air, but varies as the square root of the period, so that if two notes were sounded the high note would arrive after the low note, and the resultant effect would be entirely destroyed. Here, again, it is difficult to see how the differential equation and its solution can be evaded.

Though the history of the telegraph dates only from a little more than fifty years ago, it is ancient in comparison with the other great applications of electrical science, which have received their development during the last fifteen years. Here again mathematics which are not quite elementary have played their part. In the theory of transformers we find another illustration of the need of knowing how formulæ are obtained if they are to be correctly applied. The early transformers were made with unclosed magnetic circuits; there was an iron core, but the lines of magnetic force passed through air for a considerable part of their path. In this case a complete mathematical theory was not very difficult. But speedily closed magnetic circuits were found to be better, and the relation of magnetic induction and magnetic force became all important. If anyone were to apply mathematical formulæ, which were true for the earlier transformers, to the later ones, his results would be inaccurate. Indeed a wholly different method of attack on the problem was needed, taking account of the facts as they are, and not applying results which were true of older apparatus to cases essentially distinct*.

The employment of alternating currents has brought into use, as a necessity for understanding the actually observed phenomena, a great deal of mathematics. Why is the apparent resistance of a conductor greater for an alternating current than for a direct current? And by resistance I do not mean the quasi-resistance due to self-induction†. The mathematical electrical theory is ready with an answer; it is ready, too, to tell us how the difference depends upon the frequency of the current and on the size of the conductor. In the case of a cylindrical conductor the solution involves a knowledge of Bessel's functions. We learn that if the current has a high frequency, or if the conductor be large, there will be very little current in the centre of the cylinder, and that therefore for any practical

---

* *Proceedings of Royal Society*, February 17th, 1887.
† Lord Rayleigh, *Phil. Mag.* vol. XXI. p. 381.

purpose the centre of the cylinder might just as well not be there; the current is largely confined to the part of the conductor near to its surface. The currents at different depths in the conductor attain to their maximum values at different times; those near the surface of the cylinder occur before those at some distance from the surface. The mathematical conditions are expressed by the same equation as is used to express the disposition of heat in a cylinder the surface of which is submitted to a periodic variation of temperature. Anyone who had thoroughly mastered the heat problem would be quite prepared to deal with the problem of currents in a conductor. It cannot be too often repeated, any piece of pure mathematics which finds one application to a physical problem is almost sure to find, in exactly the same form, applications to other problems which superficially are absolutely distinct. The differential equation in this case is $kc \dfrac{dv}{dt} = \left(\dfrac{d^2v}{dr^2} + \dfrac{1}{r}\dfrac{dv}{dr}\right)$, the similarity of physical condition to the problem of linear propagation of heat is close, but the mathematics differs materially owing to the presence of the term $\dfrac{1}{r}\dfrac{dv}{dr}$ in the equation. Mathematics deals with the relation of quantities to each other without troubling as to what the physical meaning of the quantities may be. Hence it is that the mathematical treatment of two such problems as the distribution of currents in a cylindrical conductor and of heat in a cylinder is identical, whereas the treatment of the distribution of heat in a cylinder is quite distinct from the treatment of the distribution of heat in a sphere or in a solid bounded by two parallel planes.

A curious phenomenon was observed in the large alternate-current machines at Deptford when connected to the long cables intended to take the current to London. The pressure at the machines when connected to the conductors was, under certain conditions, actually greater than when not so connected. The phenomenon is one of resonance very analogous to the heavy rolling of ships when the natural period of roll is about the same as the period of the waves[*]. The period of the alternating current corresponds to the period of the waves, the self-induction of the machine to the moment of inertia of the ship, the reciprocal of the capacity to the stiffness of the ship, and the electrical resistance

[*] Institution of Electrical Engineers, November 13th, 1884.

of the conductors to the frictional resistance to rolling. The mathematics in the two cases is then the same. The effect was predicted long before it was observed in a form calculated to cause trouble.

A problem which is still agitating electrical engineers is that of running more than one alternate-circuit dynamo machine connected to the same system of mains. Before the matter became one of practical concern, it was considered in this room, and it was shown mathematically, that it was possible to run independently-driven alternators in parallel but impossible to run them in series. That is to say, that if two alternators were connected to the same mains they would tend to adjust themselves in relation to each other so that their currents could be added, but that if an attempt were made to couple them, so that their pressures should be added, they would adjust themselves so that their effects would be opposed*.

Perhaps of all engineering problems which have received their solution in the last hundred years that of the greatest practical importance is the conversion of the energy of heat into the energy of visible mechanical motion. The science of thermodynamics has advanced along with the practical improvement of the steam-engine. By its aid, particularly by the aid of the so-called second law, we know what is possible of attainment by the engineer under given conditions of temperature. I must not trench on the subject of one of my successors, but I may point out that our knowledge of the second law of thermodynamics was first developed by means of mathematics, and that to-day its neatest expression is by means of partial differential coefficients. The two most notable names in connection with the development of the second law of thermodynamics in harmony with the first are those of Kelvin and Clausius; both dealt with the subject in a mathematical form not comprehensible to those who have not had substantial mathematical training.

Illustrations such as these might be multiplied almost indefinitely. They show that the advancement of the Science of Engineering has been aided in no inconsiderable measure by the labours of mathematicians directly applying the higher mathematical methods to engineering problems. They show, too, one

---

* *Minutes of Proceedings Inst. C.E.*, April 5th, 1883; Institution of Electrical Engineers, November 13th, 1884.

way in which respect for a formula may be dangerous, one way in which it is true that mathematics may be a bad master. In St Venant's problems we have an example in which the use of older results of limited application in cases where the assumptions on which they rest are not true will mislead. The examples show the proper remedy; it is a more complete application of mathematical methods. The error is just one which a man will make who has the power to use a formula without a ready understanding of how it is arrived at. A practical man, ignoring mathematical results, might or might not escape the error of supposing that a triangular shaft would break at the angles under torsion; the half-educated mathematician would certainly fall into the snare from which complete mathematical knowledge would deliver him. You can only secure the services of that good servant, mathematics, and escape the tyranny of a bad master by thoroughly mastering the branches of mathematics you use. The mistake caused by the wrong application of mathematical formulæ is only to be cured by a more abundant supply of more powerful mathematics.

There is another drawback to the use of results, taken, it may be, out of an engineering pocket-book by those who are not prepared to understand how they are reached and on what foundations they rest. The educational advantage is lost. The close observations which enabled the earlier engineers to proportion their means to the end to be attained was no doubt very laborious, and the results could not be applied to cases much different from those which had been previously seen, but the effect on the character of the engineer was great. In like manner, to thoroughly understand the theory of an engineering problem makes a man able to understand other problems, and in addition to this precisely the same mathematical reasoning applies to many cases. The mere unintelligent use of a formula loses all this; it leaves the mind of the user unimproved, and it gives no help in dealing with questions similar in form though different in substance.

But even the use of mathematics by competent mathematicians is not without drawbacks. Mathematical treatment of any problem is always analytical—analytical, I mean, in this sense that attention is concentrated on certain facts, and other facts are neglected for the moment. For example, in dealing with the thermodynamics of a steam-engine, one dismisses from consideration very vital points essential to the successful working of the engine, questions of strength of parts, lubrication, convenience for repairs. But if an engineer is to succeed he must not fail to consider every element necessary to success; he must have a practical instinct which will tell him whether the instrument as a whole will succeed. His mind must not be only analytical, or he will be in danger of solving bits of the problems which his work presents, and of falling into fatal mistakes on points which he has omitted to consider, and which the plainest, intelligent, practical man would avoid almost without knowing it.

Again, the powers of the strongest mathematician being limited, there is a constant temptation to fit the facts to suit the mathematics, and to assume that the conclusions will have greater accuracy than the premises from which they are deduced. This is a trouble one meets with in other applications of mathematics to experimental science. In order to make the subject amenable to treatment, one finds, for example, in the science of magnetism, that it is boldly assumed that the magnetization of magnetizable material is proportional to the magnetizing force, and the ratio has a name given to it and conclusions are drawn from the assumption, but the fact is, no such proportionality exists, and all conclusions resulting from the assumption are so far invalid. Wherever possible, mathematical deductions should be frequently verified by reference to observation or experiment, for the very simple reason that they are only deductions, and the premises from which the deductions are made may be inaccurate or may be incomplete. We must always remember that we cannot get more out of the mathematical mill than we put into it, though we may get it in a form infinitely more useful for our purpose.

Engineers no doubt regard their profession from very different points of view; some think it a mere means of making money, some regard it as an instrumentality for benefiting the race, whilst others again delight in it as an interest in itself, and delight in it most of all when new knowledge is added to that which we know already. It is just the same with the medical profession; some attend patients for the guineas they receive, some give a very high place to motives of benevolence, whilst others love it as a field where new knowledge may be found and the delight of discovery enjoyed. In regard to the first class of engineers, I have no doubt a little skill in managing a board of directors or impressing a committee of Parliament will be much more useful to the engineer than a great deal of mathematics. Let him manage his board and buy his mathematician, and it is very probable he will make much more money than the mathematician or any other person of skill whom he may employ. But we cannot all of us make money in this way. In the future it is likely that educated men will have to work harder and receive less, and it is a great thing if their work can be made itself a joy, and surely this can best be by a thorough understanding of the reason of all they do, by the feeling that they have full competence to form their own judgments without depending much on the authority of others. This can only be, in the words of Sir John Herschel, by a "sound and sufficient knowledge of mathematics, the great instrument of all exact inquiry, without which no man can ever make such advances in any of the higher departments of science as can entitle him to form an independent opinion on any subject of discussion within their range."

After all, in any department of applied or pure science the highest satisfaction comes from accomplishing that which no one has done before, from disclosing what no one hitherto has known. If a department of the Arts or Sciences ceases to advance and becomes simply the application in known ways of known principles to obtain known ends, that department has lost its charm till the time comes for a fresh advent of change and development. To effect such advances it is easy to show that mathematics is a most necessary instrument. Here it is no drawback that the mind of the discoverer is too analytical; he may deal at his pleasure with one aspect of a problem, and it does not detract in any way from the value of his solution that he does not touch on incidental matters. Some of you who love the interest of continual advance in our science and practice, may look forward with a shade of sadness to a possible time when all is done or known which can be done or known, and the work of the engineer shall be merely applying principles discovered by his predecessors. In such a state, when the experience of the older generations shall control the practice of to-day, the free use of mathematical methods may be effectually superseded by the application according to rule of mathematical formulæ. But it would be a much less interesting condition than the constant change of to-day, when the practical experience of ten years ago is in many departments rendered worthless by later discoveries. But we need not fear that such a time of petrifaction will come so long as, whilst reverencing the discoverers who have added to our knowledge, we endeavour to replace their methods by better, and expect that those who come after us will, in their time, improve upon ours. Our knowledge must always be limited, but the knowable is limitless. The greater the sphere of our knowledge the greater the surface of contact with our infinite ignorance.

Mr ALFRED GILES, President, said, to those cognizant with the higher branches of mathematical science, the lecture of Dr Hopkinson would be an intellectual treat. To those not so conversant with that subject, the lecture would, he hoped, have the effect of spurring them, particularly the younger members of the profession, to further exertions to follow the brilliant example set them by Dr John Hopkinson. He was sure all present would join in offering to the lecturer their most cordial thanks for his brilliant lecture. Dr Hopkinson had taken immense pains in assembling and presenting the questions that he had brought forward. There was one subject specially mentioned, of which Mr Giles had some practical knowledge, and to which he was glad to refer—he alluded to the excellent compass invented by their Honorary Member, Lord Kelvin, better known in that connection as Sir William Thomson. He could assure the Meeting that from an intimate acquaintance with many seafaring men he had only heard one opinion as to the excellence and superiority of Sir William Thomson's compass. He was glad to have the opportunity of expressing his thanks to Lord Kelvin personally for the advantage he had conferred in that respect upon navigators.

In asking the meeting to thank the lecturer, he might observe that it was not very often that they had the opportunity of receiving good advice from a senior wrangler as on the present occasion. What had fallen from Dr Hopkinson's lips would, it might be hoped, induce the younger members of the engineering profession to exert themselves in the directions indicated. He called upon all present to join him in according most hearty thanks to Dr Hopkinson for the lecture.

The Resolution was carried by acclamation.

Dr JOHN HOPKINSON, in acknowledging the Resolution, confessed that when he undertook the lecture, he had no idea how difficult a subject it would be to deal with; and he very much feared he might have completely failed to make it at all interesting to many present. However, the members might be sure of this, that the exceedingly kind way in which they had received the attempt he had made to place the subject before them was ample reward for any trouble it had cost him.

APPENDIX.

[Extract from paper on *Magnetisation of Iron. Roy. Soc. Trans.* 1885.]

Let $\mathfrak{H}$ be the magnetic force at any point, $\mathfrak{B}$ the magnetic induction, and $\mathfrak{J}$ the magnetisation (*vide* Thomson, reprint, Maxwell, vol. ii., *Electricity and Magnetism*), then $\mathfrak{B} = \mathfrak{H} + 4\pi\mathfrak{J}$. We may therefore express any results obtained as a relation between any two of these three vectors; the most natural to select are the induction and the magnetic force, as it is these which are directly observed. $\mathfrak{B}$ is subject to the solenoidal condition, and consequently it is often possible to infer approximately its value at all points, from a knowledge of its value at one, by guessing the form of the tubes of induction. $\mathfrak{H}$ is a force having a potential, and its line integral around any closed curve must be zero if no electric currents pass through such closed curve, but is equal to $4\pi c$ if c be the total current passing through the closed curve. In arranging the apparatus for my experiments, I had other objects in view than attaining to a very small probable error in individual results. I wished to apply with ordinary means very considerable magnetising forces; also to use samples in a form easily obtained; but above all to be able to measure not only changes of induction but the actual induction at any time. The general arrangement of the experiments is shown in fig. 1, and the apparatus in which the samples are placed in fig. 2. In the latter fig. AA is a block of annealed wrought iron 457 millims. long, 165 wide, and 51 deep. A rectangular space is cut out for the magnetising coils BB. The test samples consist of two bars CC', 12·65 millims. in diameter; these are carefully turned, and slide in holes bored in the block, an accurate but loose fit; the ends which come in contact are faced true and square; a space is left between the magnetising coils BB for the exploring coil D, which is wound upon an ivory bobbin, through the eye of which one of the rods to be tested passes. The coil D is connected to

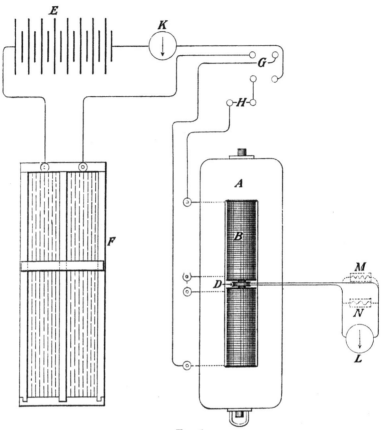

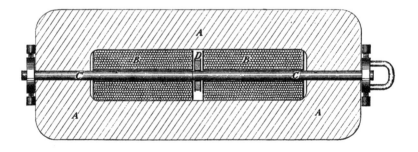

Fig. 1.

Fig. 2.

APPENDIX. 291

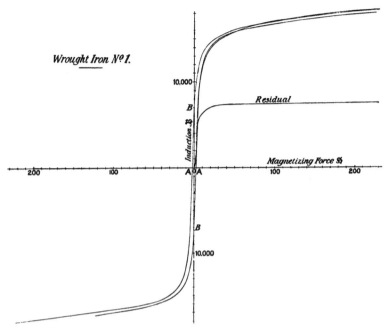

Fig. 3.

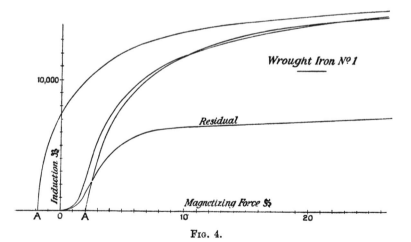

Fig. 4.

the ballistic galvanometer, and is pulled upwards by an india-rubber spring, so that when the rod C is suddenly pulled back it leaps entirely out of the field. Each of the magnetising coils B is wound with twelve layers of wire, 1·13 mm. diameter, the first four layers being separate from the outer eight, the two outer sets of eight layers are coupled parallel, and the two inner sets of four layers are in series with these and with each other. The magnetising current therefore divides between the outer and less efficient convolutions, but joins again to pass through the convolutions of smaller diameter. The effective number of convolutions in the two spools together is 2008. Referring to fig. 1, the magnetising current is generated by a battery of eight Grove cells E, its value is adjusted by a liquid rheostat F, it then passes through a reverser G, and through a contact-breaker H, where the circuit can be broken either before or at the same instant as the bar C is withdrawn; from H the current passes round the magnetising coils, and thence back through the reverser to the galvanometer K. The galvanometer K was one of those supplied by Sir W. Thomson for electric-light work, and known as the graded galvanometer, but it was fitted with a special coil to suit the work in hand. The exploring coil D was connected through a suitable key with the ballistic galvanometer L. Additional resistances M could be introduced into the circuit at pleasure, and also a shunt resistance N. With this arrangement it was possible to submit the sample to any series of magnetising forces, and at the end of the series to measure its magnetic state; for example, the current could be passed in the positive direction in the coils B, and gradually increased to a known maximum; it could then be gradually diminished by the rheostat F to a known positive value, or it could be reduced to zero; or, further, it could be reduced to zero, reversed by the reverser G, and then increased to any known negative value. At the end of the series of changes of magnetising current, the circuit is broken at H (unless the current was zero at the end of the series), and the bar C is simultaneously pulled outwards. Three successive elongations of the galvanometer L are observed. From the readings of the galvanometer K, the known number of convolutions of the coils B, and an assumed length for the sample bars, the intensity of the magnetising force $\mathfrak{H}$ is calculated. The exploring coil D had 350 convolutions. From its resistance, together with that of the galvanometer with shunts, the sensibility of the galvanometer, its time of oscillation, and its logarithmic decrement, a constant is calculated which gives the intensity of induction in the iron from the mean observed elongation of the galvanometer. The resistances have been corrected in the calculation for the error of the

APPENDIX. 293

B. A. unit, and both galvanometers were standardised on the assumption that a certain Clark's cell had an electromotive force of $1\cdot 434 \times 10^8$ C.G.S. units. This Clark's cell had been compared to and found identical with those tested by Lord Rayleigh.

Let the mean length of the lines of induction in the sample be l, and σ the section of the sample; let l' be the length of lines of induction in the block, and σ' their section, $\mathfrak{B}$ the intensity of induction in the sample, $\mathfrak{B}'$ in the block, then $\sigma\mathfrak{B} = \sigma'\mathfrak{B}' = I$; let

$$\mathfrak{B} = \mu\mathfrak{H},$$

and

$$\mathfrak{B}' = \mu'\mathfrak{H};$$

then

$$\int \left(\frac{\mathfrak{B}}{\mu} + \frac{\mathfrak{B}'}{\mu'}\right) ds = \int \mathfrak{H} ds$$

or

$$\frac{Il}{\mu\sigma} = 4\pi nc - \frac{Il'}{\mu'\sigma'},$$

where n is the number of convolutions of the magnetising coils. Now in the instrument used σ' is large, and μ' is as large as can be obtained, hence the term $\frac{Il'}{\mu'\sigma'}$ is small comparatively. My first intention was to correct the magnetising force by deducting this small correction, but finally I did not do so, because in the more interesting results the magnetism of the block is dependent in part upon previous magnetising forces, the effect of which cannot be allowed for with certainty. We know then that in all the curves the magnetising force indicated is actually too great by a small but sensible amount, which does not affect the general character of the results or their application to any practical purpose. The magnetising force then at any point of the sample is $\frac{I}{\sigma\mu} = \frac{4\pi nc}{l}$ – a small correction which we deliberately neglect. There is another source of uncertainty in the magnetising force: the length l is certainly greater than the space within the wrought-iron block, but it is not possible to say precisely how much greater. If the sample bars and the block were a single piece, the results of Lord Rayleigh for the resistance of a wire soldered into a block would be fairly applicable; but it is essential that there should be sufficient freedom for the bar to slide in the hole; the minute difference between the diameters of the sample and the hole will increase the value which should be assigned to l. Throughout, l is assumed to be 32 centims., and it is not likely that this value is incorrect so much as half the radius of the bar, or 1 per cent. The magnetising forces ranged up to 240 c.g.s. units when both bars were of the same material. In some cases a single bar only was available for experiment; the plan then was to use it as the bar which enters into the exploring coil, and for the other to use a known bar of soft iron. We have then to deduct from $4\pi nc$ the magnetising force required to magnetise the bar of soft iron to the state observed, and to distribute the remainder over the shorter length of sample examined. The results obtained in this way are subject to a greater error, because some lines of induction undoubtedly make their way across from the end of the soft-iron bar to the body of the block. A small correction is required, important in the case of bodies but slightly magnetic, for the fact that the area of the exploring coil is greater than the area of the bars tested. Thus the induction measured by the exploring coil is not only that in the sample, but something also in the air around the sample. The amount of this was tested by substituting for a sample of iron or steel a bar of copper, and afterwards a rod of wood, and it was found in both cases that the induction $\mathfrak{B}$ was 370 when the force $\mathfrak{H}$ was 230. The correction is in all cases small, but it has been applied in the column giving the maximum induction, as it materially affects the result when the sample contains much manganese, and is consequently very little magnetic.